办公空间创意设计

FROM ME TO WE
THE CHANGING WORKPLACE

（英）马修·德里斯科尔／编
常文心／译

辽宁科学技术出版社

PREFACE
前言

THE CHANGING NATURE OF WORKSPACE DESIGN

The contemporary workspace is beginning to change and that change has been led by a new wave of creative and technology businesses. The office has for long been not so different from its mid twentieth century iterations. Whilst the cubicle has gone in preference for groups of facing workstations, the large open plan floor plate filled with grids of desks endures as the common design for much of the corporate world. Whilst the tools of the trades have changed enormously along with the building envelopes which shelter the staff, most people remain the occupants of a fixed 2mx2m floor plate with a chair and desk. Why has this compartment endured so long? Economics, technology, practicality? It is certain that a high person to area ratio keeps cost down and profits up, and until relatively recently even the technology required an umbilical supply of power and data. What are the advantages to the individual afforded by the workstation compartment (with or without walls)? People like a personalised domain, a space they can make their own and even more so within a homogenous environment. There are productivity benefits to grouping people in teams, but conversely this stifles cross fertilisation and wider networking within an organisation.

Sadly this kind of environment comprising of workstation clusters, office pods, meeting rooms and occasional breakout proliferate the world, but this is changing. At the forefront of this change are the new wave of media, tech and creative businesses. These young dynamic and growing companies are unburdened by historic company structures and working practices, and critically they have embraced technology which allows total mobility and flexibility. Wireless technology and cloud-based data have unleashed the worker from their desk and enabled the design of the workplace to change. This physical freedom has enabled greater flexibility in the constraints of the working day. The boundaries between personal and professional life have become increasingly blurred and modern business has recognised that the provision of a more stimulating and comfortable environment enhances productivity.

The first iteration of flexible space design was "hot desking" which encouraged workers to choose their location with each workstation providing access to the tools they need. However, this still tended to be only the freedom to choose what was ostensibly a traditional desk. In offices where this is practiced it is often found that people would choose a station and then return to the same place each day; this is a desire for a personal domain, a sense of individuality, the need for a "home". Perhaps, but equally in the absence of variety it seems logical that one would settle into a routine, and there is comfort in the known. If the only variety of choice in the workspace is the group of people inhabiting the space around you then the tendency to build a community is normal. Does this make hot desking a failure, in this outcome it certainly fails to foster movement, cross communication and chance encounter which must be at the heart of the idea. However, perhaps the failure is to not recognise the importance of variety and choice of environment, and this is the direction that the contemporary workspace is moving.

The mobility of the worker is critical to this allowing staff complete freedom within

办公空间设计的变化

随着创新企业和技术企业的浪潮，现代办公空间正在发生变化。办公室已经不再是20世纪中期的模样了。虽然小隔间已经被面对面的群组工作台所取代，但是开放的大楼板上办公桌仍然纵横交错。无论是交易工具，还是建筑外观都经历了巨大的变化，而大多数人仍被困在一个2米×2米的空间内。为什么隔间设计能坚持这么久？经济原因，技术原因，还是实用原因？当然，人均面积越少，办公成本越低，利润越高。而且一直以来，技术仍然要求集中的电力和数据供应。那么，隔间式办公台（无论是否有隔断墙）对个人的好处是什么呢？人们喜欢私人化的领域，一个能够释放自我的空间。将人们集中起来确实能提高生产力，但是相反的，这也会影响跨界交流和更广泛的沟通。

这种由工作组、独立办公室、会议室和临时休息空间所构成的办公环境遍布全球。令人欣慰的是，这种状况正在发生变化，走在前列的媒体、技术和创意产业，这些年轻而充满活力的公司摆脱了传统的公司结构和办公方式，接受了具有高自由度和高灵活性的新技术。无线技术和云数据把员工们从办公桌前解放出来，为办公空间的改变提供了可行性。身体上的自由进一步实现了更好的灵活性，私人生活和职业生活的界限越来越模糊，现代商务已经认可了更具启发性、更舒适的环境能提高生产效率。

第一代的灵活空间设计是"办公桌轮用制"，这种设计鼓励员工根据所需的工具来选择办公桌，但是他们所选择的办公桌在外观上也就是传统的办公桌。在采用这种制度的办公室，人们通常会选择一个办公桌，然后坐上一整天，因为人类总是渴望占有独立空间，有"家"的需求。看起来，人们更倾向于固定的模式，这让他们更舒服。在办公空间里，如果只能与他人聚在一起，那么有必要建立一种社区模式。"办公桌轮用制"的失败在于它无法促进移动，无法促进跨界交流，也无法提供邂逅机遇。也可能，它的失败应当归咎于没有认识到多样化环境的重要性，而这种多样化选择正式现代办公空间的发展方向。员工的移动性是赋予员工在办公室内完全自由的关键，设计师应当为不同的工作任务设计出多种多样的办公环境。这种灵活的办公空间和完全的自由并不适合所有商务工作，但是每个行业都应尽量开发探索。这种思路已经彻底改变了办公空间的设计前景，让设计师能够自由想象办公空间的模样。科技的进步实现了笔记本电脑的普遍化，使其代替了台式电脑、电话和笔记本。这种变化意味着随处都能成为一个工作站，无论是公园的长椅还是树枝。这种移动性模糊了办公环境内工作与社交空间的界限。咖啡厅、会议室、集体空间都能成为工作站，传统办公环境被分割开来，更轻松，也更多样。越来越多的现代办公空间给人以家一般的轻松感，营造出舒适却具有启发性的环境。

the office which permits the designer to create varied workspaces environment with different spaces and conditions for different tasks. This kind of agile workspace and total freedom is not suited to all business practices, but every business can exploit it to some extent. The affect this has on workspace design is enormous completely changing the landscape of potential allowing the designer the freedom to reimagine what an office can be. Technological advancement has turned the laptop into the complete tool for many of us, replacing the desktop computer, the telephone, and the notebook. This shift means that almost anything can be a workstation, from a park bench to the branch of a tree. This mobility enables the blurring of boundaries between work and social space within the office environment. Now the café can function as a workstation, meeting room or collaborative space, and this allows the traditional workspace environment to break up and become relaxed and varied. Increasingly contemporary workspaces are taking on a domestic looser feel to create comfortable yet inspirational environments.

Employers recognise the need for better amenities and are putting staff health and happiness at the heart of the workplace, ranging from running tracks on the roof to in-office spa facilities and massage. This shift to centralise the wellbeing of the staff is not only because it has a proven impact on productivity but because the modern tech and design literate employees to expect their work environment to be of a certain standard and the competition for the best staff pushes business to create the best workplace. Many companies have recognised the huge importance of diet with some prioritising healthy eating. Not only is healthy eating key but the importance of sharing a meal with your colleagues and friends is central to human nature and valuing this has a significant impact on the team atmosphere. Truly innovative workspace design must go beyond colour schemes and soft furnishings and really push the boundaries of the work space and working practices exploiting technology to its full potential. As a design practice we have explored this at both ends of the spectrum creating a workspace where traditional design is counterbalanced with a circulation and breakout environment which encourages crossover and provides varied spaces for both social and professional interaction, ranging from the solo booth to the 50 people gathering. This landscape annexes the traditional workspace and acts as an "other" environment which people can visit and inhabit which encourages movement and a shift in surroundings. At the other end of the spectrum is a workspace design we have developed based on complete freedom, where no one has a desk, and there are very few traditional workstations to inhabit. Each individual has a "landing" point or home, where their personal effects are located, a place at which to start the day. After that a wide range of workspace conditions which can be inhabited to suit the tasks they are carrying out range from collaborative project spaces, shared desks, relaxed seating and solo booths. This kind of freedom enables people to find the right place to carry out a particular task and encourages movement, cross over and collaboration. The end of the traditional workstation is in sight, to be replaced with an open free environment of mobile working.

This book shows a range of projects which have explored what a contemporary workspace can be and demonstrate the variety and creativity present in modern workplace design.

目前，雇主已经认识到了康乐设施的必要性，他们将员工的健康和幸福放在办公空间的核心位置，设置了屋顶跑道、办公室SPA设施、按摩房等。这种转变不仅因为员工的身心健康对生产效率有重大影响，还因为现代技术和设计文化让员工们对自己的办公环境有了某种标准的要求。为了赢得最好的员工，公司必须提供最好的办公空间。许多公司还认识到了饮食的重要性，把健康饮食放在了首位。事实上，重点不仅在于健康饮食，更在于与同事朋友共进午餐的过程，这是人类的天性，也是团队建设的重要组成部分。这种有创意的办公空间设计必须超越简单的色彩搭配和软装饰，应当充分利用科技来拓展办公空间和办公实践的界限。

作为一家设计公司，我们对两方面都进行了探索，由此打造出融合的办公空间。一方面，我们在传统设计中加入了流通和休息环境，鼓励跨部门交流，同时也提供了独立包间、50人聚会空间等各种类型的社交空间和职业互动空间。这种新增的空间对传统办公空间进行了补充，鼓励人们在其间穿梭活动。另一方面，我们也开发了完全自由的办公空间。办公室里没有办公桌，传统式的工作站也极少。每个人都有一个"着陆点"或"家"，那里放置着他们的私人物品，是他们每天工作的起点。合作项目空间、共享办公桌、休闲座位和独立包间等办公环境能满足各种各样的日常任务。这种自由度让人们能找到执行特定任务的正确地点，并且能鼓励移动、交流和协作。传统的工作台即将被开放自由的移动办公环境所取代。

本书将带你探索现代办公空间的各种可能性，领略各种各样的创新办公设计的魅力。

ABOUT THE AUTHOR
作者简介

Matthew Driscoll co-founded Threefold Architects with Jack Hosea in 2004. Since setting up practice he has been key to the development and success of Threefold Architects. He is a lead designer and as well as acting as project director to deliver many projects including Ladderstile House, Wharf Green and Hurst Avenue.
Matthew was educated at the Bartlett, UCL. Following his graduation he worked for both large and small award winning design practices where gained a variety of experience working on new building projects from houseboats and bridges to racecourses and office buildings. Matthew's experience has focused on innovation in design and use of materials and construction techniques.

2004年，马修·德里斯科尔与杰克·霍齐亚共同创建了Threefold建筑事务所。自从开业以来，他对事务所的发展做出了巨大贡献。作为主创设计师和项目总监，他完成了拉德斯蒂尔住宅、格林码头、沙洲大道等多个项目。
马修毕业于伦敦大学巴特列特建筑学院。毕业后，他曾在大大小小的优秀建筑事务所中工作，获得了船屋、桥梁、赛马场、办公楼等多方面的项目设计经验。马修的经验主要集中在设计、材料和建造技术的创新应用上。

CONTENTS
目录

008 PART 1 COLLABORATION - FOR IDEA CONVERGENCE AND TEAM WORK
第1章 协作型办公空间——有利于思想的交汇与群体工作

014 FUZZY ZONE AND STAFF'S COLLABORATION
模糊区域的建立与员工间的协作
——*PIXIV OFFICE*
　PIVIX 公司

020 SPATIAL COLOURS REFLECT TEAM SPIRITS
空间色彩体现群体的合作精神
——*APOS^2*
　艾普斯平方

028 A RIBBON-LIKE JOINERY SYSTEM HELPS INTERDEPARTMENTAL COLLABORATION
带状的工作台有利于部门间的协同
——*CHEIL HONG KONG*
　香港杰尔思行广告公司

034 AN OPEN LAYOUT PROMOTES TEAM WORK
开放的布局方式促进团队工作
——*LE CAMPUS*
　校园办公

040 AUXILIARY SPACES ENCOURAGE DIVERSIFIED TEAM INTERACTIONS
辅助空间可促成团队互动的多样化
——*LINKEDIN 605 MAUDE SUNNYVALE*
　LinkedIn 领英公司森尼韦尔市莫德街605号办公楼

048 THE FLEXIBLE USE OF COLOURS REFLECTS THE TEAM'S CORE VALUE
色彩的灵活运用可体现集体的核心价值
——*NTI OFFICE*
　NTI 办公室

054 APPROPRIATE ZONING IN LINE WITH THE TEAM'S CREATION PROCESS
根据团队创意流程进行合理的分区
——*SPACES ARCHITECT'S @KA OFFICE*
　Ka 建筑事务所

064 PART 2 EXCHANGE - FOR KNOWLEDGE'S EXCHANGE, SHARING AND INCREASE
第 2 章 交流型办公空间——有利于知识的交换、共享和增长

070 THE EXPERIENTIAL SPACE ENCOURAGES IDEA EXCHANGES
体验功能的空间有助于思想的互动
——ONE WORKPLACE HEADQUARTERS
第一车间公司总部

080 A HIGHLY FLEXIBLE WORKSPACE TAPS THE TEAM'S POTENTIALS
高度灵活的空间有利于团队潜力的发挥
——OFFICE MINDMATTERS
Mindmatters 公司办公室

086 DIVERSIFIED FACILITIES PROMOTE STAFF'S COMMUNICATIONS
变化多样的设施促进员工思想的交流
——HEAVYBIT INDUSTRIES
重比特实业

098 AN OPEN WORK ENVIRONMENT HELPS INDIVIDUALS EXPRESS THEIR CHARACTERISTICS
开放的空间氛围激励员工个体特色的表达
——TRIBAL DDB AMSTERDAM
阿姆斯特丹 Tribal DDB 公司

104 VARIOUS SOCIAL SPACES SATISFY STAFF'S INTERACTIVE REQUIREMENTS
多样的社交空间满足员工互动的需求
——BRIGHTLANDS CHEMELOT, BUILDING 24
切美洛特化工园 24 号楼

112 DECORATIONS WITH VARIOUS COLOURS CREATE A RELAXING COMMUNICATION ATMOSPHERE
通过不同色彩的装饰营造轻松的沟通氛围
——GOOGLE TOKYO
谷歌东京

118 EFFECTIVE CONNECTIONS BETWEEN AREAS FACILITATE INTERDISCIPLINARY COMMUNICATIONS
各区域的有效连通促进跨领域沟通
——KOIL – KASHIWA-NO-HA OPEN INNOVATION LAB
柏市开放创新实验室

126 THE BREAK OF CLOSED OFFICE AND THE FREE EXCHANGES BETWEEN EMPLOYEES
封闭办公区域的打破与员工的无障碍交流
——MOZILLA VANCOUVER OFFICE RENOVATION
摩斯拉公司温哥华办公室

134 PART 3 COMMUNITY - GREAT INCLUSIVITY TO SPACES COMBINING WORK AND SOCIAL ACTIVITIES
第 3 章 社区型办公空间——对工作与社交相结合的场所有极大包容性

140 CULTURAL THEMES CREATE MULTI-FUNCTIONAL WORKSPACE
以文化特色为主题打造出多功能办公空间
—— *GOOGLE ISRAEL OFFICE TEL AVIV*
谷歌特拉维夫办公室

148 THE NON-PARTITION RELAXING CENTRE CREATES AN URBAN LIFESTYLE
无区隔的休闲区营造城市生活的氛围
—— *COMCAST SILICON VALLEY INNOVATION CENTRE*
康卡斯特硅谷创新中心

156 CREATING A HOME-LIKE ATMOSPHERE WITH "VILLAGE" AS ITS CORE ELEMENT
以"村庄"为核心元素营造居家办公的氛围
—— *HALLE A*
A 厅办公空间

166 PART 4 MOBILITY - SETTING OF NON-REGIONAL WORKING BELT
第 4 章 流动型办公空间——非区域性工作带的设置

170 SCATTERED BOXES ENABLE FREE WORKSTYLE
散落的盒子令自由的工作形式成为可能
—— *1305 STUDIO*
1305 工作室

182 USING A "BRIDGE" TO CONNECT DIFFERENT AREAS
以"桥梁"作为各区域的连接元素
—— *THE BRIDGE*
桥梁办公

192 THE MOBILE FURNITURE ENABLES CHANGEABLE SPACE
设施的可移动性打造空间的可变性
—— *ORANGE GROVE ATHENS*
雅典橘林产业园

200 INTEGRATION WITH URBAN PUBLIC SPACES
城市公共空间的巧妙植入
—— *JUSTPEOPLE*
唯人公司

208 FREE CHOICES BETWEEN COMPARTMENTS AND COMFORTABLE WORKSPACE
格子间与舒适办公的自由选择

——OoO PRESS2
　PRESS2 办公空间

214　WORKSPACES DEFINED WITH SPECIFIC MISSIONS
　　　由具体工作选择不同的办公场所
　　　　——FRIENDS OF EARTH
　　　　　地球之友

224　DIFFERENT WORKSPACES FACILITATE DIFFERENT THINKING PROCESSES
　　　通过不同的办公环境辅助不同的思维过程
　　　　——SKYPE'S NORTH AMERICAN HEADQUARTERS
　　　　　Skype 公司北美总部

234　PART 5　SMALL AND SMART - HOW TO MAKE FULL USE OF WORKSPACE
　　　第 5 章 集约型办公空间——如何实现"空间尽其用"

240　THE CONTRAST BETWEEN DARK GREY AND WHITE ENLARGES THE SPACE VISUALLY
　　　深灰与纯白的对比产生了放大空间的效果
　　　　——ARCHITECTURE STUDIO AND COWORKING SPACE OFFICE
　　　　　建筑工作室与协作办公空间

248　FLUID FORMS AMPLIFY THE SPACE
　　　流线的运用产生了扩大空间的作用
　　　　——FLUID FORMS – CUBIC OFFICE
　　　　　流动造型——立方办公室

254　THE SMART USE OF PLANTS PROMOTES THE OFFICE'S QUALITY
　　　巧妙利用植栽提升办公空间品质
　　　　——SISII SHOWROOM AND OFFICE IN KOBE
　　　　　Sisii 公司神户展示厅兼办公室

262　THE SLOPING ROOF AND OPENINGS OVERCOME THE LIMITS OF SMALL SPACE
　　　斜屋顶与开窗克服小空间的局限性
　　　　——GOLDEN RATIO HEADQUARTERS
　　　　　黄金比例总部

268　AFTERWORD
　　　后记

270　INDEX
　　　索引

PART 1

COLLABORATION – FOR IDEA CONVERGENCE AND TEAM WORK

第1章 协作型办公空间——
有利于思想的交汇与群体工作

COLLABORATIVE INTERDISCIPLINARY SPACE FOR CREATIVITY

The latest trend of designing collaborative office space is that architects collaborate with digital makers to design. The design team is an "Ultra-technologists" group made up of various specialists, including: programme engineers, mathematicians, architects, designers, animators, and artists. Amongst a group that uses their heads to make things and their hands to create, and architects still perform the major role of design. The design team makes ambiguous the coexistence of digital and architecture. By taking digital, which up until now existed only inside the screen, and creating architecture in which space can be experienced, architects produce architectural design in which the coexistence of the digital and real are ambiguous.

Nowadays many architects are aiming to create architectural space for the Internet age. To explain, just take a cell phone as an example. A cell phone before the development of the Internet was designed solely for the purpose of making telephone calls. As a result the screen was small and the keyboard consisted of large physical buttons. With the development of the Internet, however, the cell phone is no longer just for calls; it is used to browse the Web. For that purpose the cell phone buttons are no longer physical buttons, and the entire screen has become a touch panel. What is the difference? Previously it was important for cell phones to have physical buttons, so they were designed with physical design as utmost important, and as the physical product was more important, little attention was paid to the screen design. The result is that no one remembers the screen design.

After the development of the Internet, not just the physical product, but the screen design has become important. The movement within the screen, and design of the screen, are part of today's cell phone design. If an iPhone has a different screen design, it becomes a different product. The movement of the icons and the design have become a part of the product.

It is obvious that the same process can occur with space design. With the information on the Internet, one uses a mouse and a browser to navigate what is displayed on screen. However, in place of the mouse it is possible to use furniture or a wall, or the space itself. For example, touch the wall and the space changes. Designers use digital information to expand space, and create works in which it

跨学科协作型创意办公空间设计

协作型办公空间设计的最新潮流是建筑师与数码设计人员协同合作。设计团队是由各种专业人士组成的"超技术团队",包括程序工程师、数学家、建筑师、设计师、动画师和艺术家。在这个团队中,建筑师仍占设计的主导地位。设计团队在数码和建筑中找到了共存的临界点。屏幕中的数码技术和可以切实体验的建筑让建筑师的建筑设计实现了虚拟与现实的共存。

当前,很多建筑师都希望打造互联网时代的建筑空间。以手机为例,在互联网发展之前,手机只有单一的打电话功能。因此,手机的屏幕很小,键盘全部由大型物理按键组成。随着互联网的发展,手机的作用已经不仅是打电话,还可以用来浏览网页,因此,手机键盘不再是物理按键,整个屏幕都成了触摸板。区别是什么呢?以前,手机的物理按键很重要,因此,物理设计是最重要的,由于物理产品更重要,所以设计师对屏幕设计也就没有那么在意。结果是没有人还记得屏幕设计。

在互联网发展之后,不仅是物理产品,屏幕设计也变得重要起来。屏幕的运动、屏幕的设计已经成为了当今手机设计的重要部分。如果苹果手机有一个不同的屏幕设计,它就变成了完全不同的产品。图标的移动和设计也成了产品设计的一部分。

很明显,空间设计也将面临同样的转变。在互联网信息时代,人们利用鼠标和浏览器来控制屏幕上显示的内容。然而,鼠标的位置也可以在家具上、墙壁上或者空间本身。例如,当你触摸墙壁时,整个空间都变化了。设计师利用数码信息来拓展空间,让人们可以体验空间。如果打造一个空间与数字可并行不悖的状态,那么空间就可以被体验。

所以,在信息社会中进行创造时,建筑师就必须考虑到空间的需求。随着互联网的出现,信息分布的层次、流量和速度都得到了巨大的提升。制造一件产品不可能只依靠一位专业人士。我们需要跨领域的交流,正如上文所提到的,我们的

becomes possible to experience the space. By creating a state in which the space and digital can proceed concurrently it is possible to create a space that can be experienced.

So, when making something in the information society, it is the architect's job to consider what is necessary in terms of the space. With the appearance of the Internet, the level, flow and speed of information distribution has increased enormously. It is no longer possible for a specialist in only one field to make a product. Communication is needed across specialties, and, as mentioned above, the structure of our organisation consists of a group of specialists in different fields, and boundaries among specialities become blurred when working towards a product. In office design as well, it has become important to design a space that activates communication between different specialists.

Creativity, communication, tension, intelligence, productivity, subjectivity and chance

In an information society we think that, creativity, communication, tension, intelligence, productivity, subjectivity, and chance, are important elements to generate in the office designs that we create.

Take Pixiv's office design as an example. The Pixiv site managed by Pixiv Co., Ltd. (http://www.Pixiv.net/) is an SNS site that specialises in providing a service for users to post drawings and illustrations. The idea behind Pixiv is to build a platform for everyone to enjoy creating pictures, regardless of whether they are good at drawing or not. Approximately 120 employees work in the offices on the ground floor of a tenant building in the heart of Tokyo.

In order to realise a Pixiv office where knowledge can be shared between employees, the walls were reduced as much as possible to allow sharing of information. There is no individual storage space, which means that each individual's belongings are placed on the desk. Like a white cube space with nothing in it, rather than a space overflowing with things, a space where anything can be done and that facilitates subjective opinions.

In addition, an intentionally large number of colours were used throughout the entire office. The colours have a psychological aspect. It is believed that by combining colours that fit the mood you wish to create, the way of conversation

组织结构由不同领域的专业人士构成，在努力制作一件产品时，专业的界线将变得模糊。在办公空间设计中，设计出一个能促进不同专业之间交流的空间变得至关重要。

创意、交流、张力、智能、生产力、主观性和机会

在这个信息社会，我们认为创意、交流、张力、智能、生产力、主观性和机会是办公空间设计的主要元素。

以 Pixiv 公司的办公空间设计为例。由 Pixiv 管理的 Pixiv 网站（http://www.Pixiv.net/）是一个社交网站，专为用户提供贴图服务。Pixiv 背后的理念是打造一个人人都可以享受创造图片的平台，无论他们擅不擅长绘画。目前，公司有约 120 名员工，他们在东京市中心的一座大厦的一层办公。

为了给 Pixiv 公司打造一个便于员工共享知识的空间，设计师尽量减少了墙壁隔断，方便信息共享。公司里没有个人存储空间，即员工的私人物品都放在办公桌上。整个办公空间就像一个空白的立方体，与充斥着各种事物的空间不同，这里空空如也，可以做任何事情，全凭个人主观意愿。

此外，整个办公空间的设计特别选择了大量的色彩。色彩具有心理效果，混合适应心情的色彩能改变对话的方式。每个场所都对应了代表情绪的色彩。例如，红色或橙色等具有积极心理效果的色彩被应用在鼓励交流的区域；而其他需要心绪平和的区域则采用了绿色。

Pixiv 办公室的标志性设计——变形虫办公桌

如果 120 人在同一层办公，他们将倾向于朝向同一个方向，而视线则会变得单调。变形虫办公桌的曲线让坐在办公桌前的视线变得多样化。

如果视线面对同一方向，对坐的两个人将会感到更焦虑，因此你需要让座位错开。然而，变形虫办公桌可以分散视线，即使两人坐的很近。这有助于集中精力，无需再去在意他人。此外，虽然在物理上是封闭的，当人们想要交谈时，却可以马上轻松地交谈。

changes. The colours become a mood that you would like to create that is adopted at each location. For example, we use a psychologically positive colour, such as red or orange, in the location where communication is to be activated. For the rest of the space, to psychologically settle the mind, green is used.

How was the symbolic amoeba desk of the Pixiv office born?

If 120 people are on a single floor, they all tend to orient the environment in the same direction, and the line of sight becomes monotonous. By curving the amoeba work desk, it becomes possible to vary the line of sight when you are sitting at the desk.

If the line of sight is facing in the same direction, people in the opposite seats will become more anxious, so you need to offset the seat to some extent. However, with the amoeba desk it is possible to disperse the line of sight, even with close seats, and it is possible to concentrate more on work without worrying about others. Yet, being physically close, when people want to talk, they can talk immediately and casually.

When the number of employees and scale of the company increases, since the division of roles and seats is gradually carried out, the feeling of being all one company can fade. With one amoeba desk, however, you can all use the same desk and feel connected. Even doing different work, employees can have greater feeling of working together.

How information is accumulated – conversation with the user

The designers actually talked with users who are active on Pixiv. What they had in common was that, to many people, the most important thing was not showing off their own picture. There were many people who said, "That picture is not good." What Pixiv is aiming for is the enjoyment of drawing. "Anyone can have fun drawing" and that is why such a mechanism is incorporated into the office design.

For this purpose in the office entrance, anybody who visits the office can feel free to enjoy drawing on 3,000 ema plaques that are installed in the walls. Ema are plaques on which Japanese draw pictures to make prayer wishes at shrines. It is a platform on which anyone regardless of their drawing expertise can draw what they feel as an offering. In Pixiv office, because anyone can freely draw on the

如果员工人数增加或公司规模扩大，工作角色和座位的逐渐分化会淡化同一公司的凝聚感。变形虫办公桌让大家坐在一起，感觉联系更加紧密。即使做着不同的工作，员工的集体感也会更强。

如何采集信息——与用户对话

设计师与Pixiv网站上活跃的用户进行了交谈。他们的共同点在于：对许多人来说，最重要的并不是炫耀自己的画作。许多人都说："那幅画不太好。"Pixiv的真正目标是享受绘画的过程。"任何人都能快乐地绘画"，这也是办公设计的出发点。

因此，在办公室的入口处，任何来访者都可以在墙壁上的3,000块饰板上随意绘画。这些饰板与日本人在寺庙中祈福用的牌子一样。这是一个让任何人都可以随意绘画的平台。在Pixiv办公室，人人都可以在饰板上自由绘画，无论是著名的画家还是学校的孩童。

最后

在未来，我们认为一切的重点都是用言语难以表达的。在信息社会中，文字的作用是快速传播、共享，快速复制，因此，它们已经失去了价值。不能轻易用文字理解的东西有抽象的价值，这些东西难以名状，却是未来的价值所在。因此，必须打造一个能让难以言喻的东西生长的框架结构，让不同领域的人聚集起来共同思考是十分重要的。一开始，看似可能浪费了大量的信息，但是，在一个能聚集不同领域人士来制作产品的空间，思想将会交汇和碰撞，这对未来的社会是极为必要的。

ema plaques, world-famous artists and schoolchildren alike can leave drawings.

Lastly

In regard to the future, what we believe is important is that which cannot be expressed in words. What words can do in an information society is to spread and share information at speed, quickly copied, and so for a business they lose value. There is abstract value in what cannot be easily understood in words, what is good but cannot be stated clearly, and in this abstract value there is a hint for the future. For that reason it is important to create a structure that gives birth to something that cannot be put into words. And, it is important to have a place where people of different specialities can think together. At first, it would seem that vast amounts of information are being wasted, but, a place that brings together different specialists to work on a product, where ideas can transverse and cross over, will become necessary for the future of society.

PIXIV OFFICE

FUZZY ZONE AND STAFF'S COLLABORATION

Tokyo, Japan

模糊区域的建立与员工间的协作
PIVIX 公司 / 日本、东京

In an information society, production process is created by the discussion of each other while showing a mock-up and moving the hand. Therefore, it is necessary to provide a space to take a vague communication in the workspace.

It does not have a clear area of personal space. To build up the area somehow ambiguous, employees will build the space by themselves. This ambiguous area between seats leads to a pretty good meeting started by accidental conversation. Because people who walk in the path will see the face of a lot of people (there are only two entrances to enter inside area of work desk), they are easy to say "Hey, what do you think about it?"

In the Pixiv office, the workplace centres around a 250m long desk that snakes throughout the entire space, zigzaging around the room and meandering around corners and against walls. Because the wooden table is completely connected, teamLab has molded its shape so that a bridge-like underpass allows workers to walk beneath it, directly through the desk. Cut-out holes act as private areas that

Pixiv's offices are created from the very art that is drawn by their guests. The walls of the entryway are composed of 3,000 ema placards. Pixiv's mission is to make it so that anyone can enjoy the fun of drawing – so anyone who visits the office can make their own picture, and the design of the entryway allows those pictures to be easily displayed in the office. The placards are affixed with magnets, so it's simple to take them down, draw another picture, and put them back up on the walls. To paint on canvas takes a certain degree of confidence in one's skills, but anyone can draw on a placard and display their work.

PIVIX 公司的办公室里装饰着由访客绘制的各种艺术画作。入口通道的两侧墙壁由 3,000 块饰板构成。PIVIX 的目标是让人人都享受绘画的乐趣，因此每个到访者都能绘制自己的图画，而入口通道的设计则让这些图画可以展示出来。饰板由磁铁固定，可以很简单地拿下来、画一幅画，然后再贴上去。在油画布上作画可能会让不自信的人却步，但是人人都可以在饰板上作画并展示出来。

individuals can climb into and remotely work with a laptop, while a three-tiered bookshelf can be ascended and allows a lofty place to settle above the rest of the facility.

Design Company: teamLab Architects   Designer: KAWATA Shogo   Client: Pixiv Inc.   Completion Date: 2013   Photography: TADA (YUKAI)   Floor: Carpet   Wall: Painted PB   Ceiling: PB

在信息社会，人们在相互讨论和交流中实现生产制作流程。因此，在工作空间中提供一个能进行交流的空间显得至关重要。这个办公空间并没有明确界定的私人空间。为了打造模糊的空间界限，员工们需要自己建立私人空间。座位之间的模糊区域能在不经意之间实现良好的对话。当一个人通过走道时，他会面对许多张脸（只有两个入口可以通往工作台的内部），因此很容易就会产生这种对话："你觉得这个怎么样？"

在PIVIX公司的办公室，一张250米长的工作台将整个办公空间环绕起来。它曲曲折折，绕过了转角，紧贴着墙壁。由于木桌是完全相连的，teamLab通过造型实现了一个拱门，使员工可以在它的下方通过。桌子上有一些切割出来的圆洞，人们可以爬进洞里，用笔记本电脑独自动作。员工还可以攀爬到三层高的书架上，在高处工作。

设计公司：teamLab建筑事务所   设计师：川田章吾   委托方：PIVIX公司   竣工时间：2013年   摄影：TADA（YUKAI）   地面：地毯   墙壁：PB涂料   天花板：PB涂料

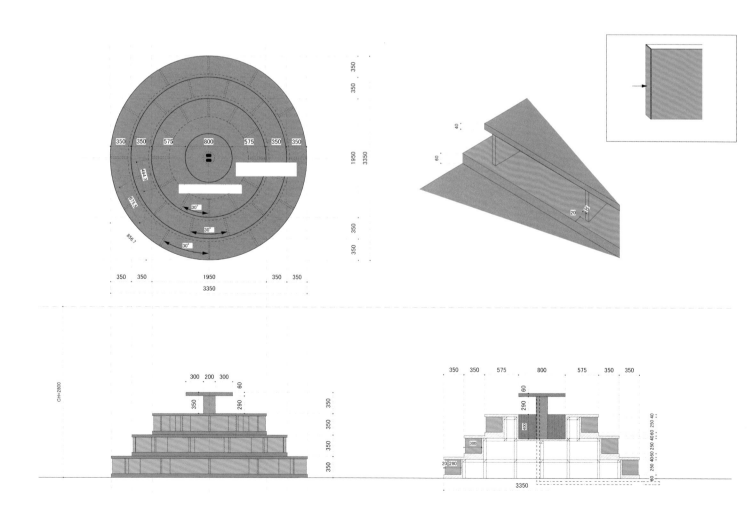

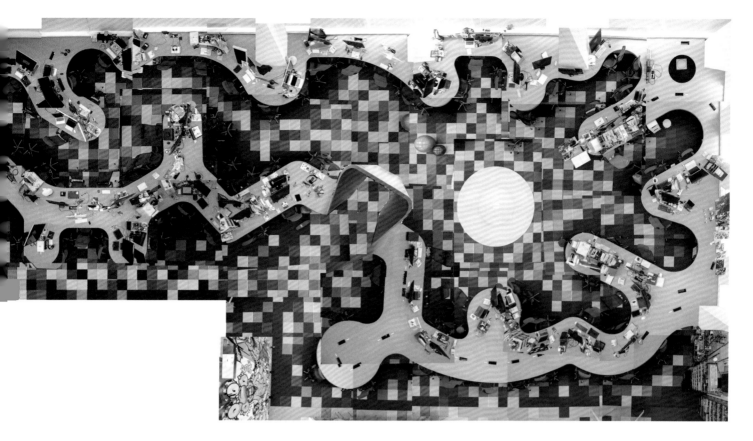

The designers aimed at an office environment where employees can work while having a conversation everywhere. The 250m work desk has no physical partition. There is kind of an iconic table in the middle inside the 250m work desk. It seems like no secret meeting for workers. It seems like a stage for meeting. Workers called it "Dohyo" (it means sumo ring).

"Dohyo" means the ring to play Sumo. Dohyo has sanctity for Japanese and it provides a feeling of tension to people. This tension can be good for the meeting. Dohyo is built in the middle of the working space. The meeting space in working space may make us be careful around. But "Dohyo" is a bit higher than other people, so it is easier to focus on the meeting. People use this meeting space for "Short" and "Focus".

设计师旨在打造一个让员工们边工作边会话的空间。250米的工作台上没有任何物理隔断。250米工作台的中央有一个标志性的圆桌，它是员工们的会面空间，让员工之间没有秘密。它的中央就像一个舞台，员工们亲切的将其称为"相扑赛场"。

"相扑赛场"即相扑的比赛场地，它在日本人的眼里是神圣而不可侵犯的，会给人一种紧张感。这与这一空间有着异曲同工之妙。办公空间里的会面空间通常会让人感到小心紧张，但是"相扑赛场"的位置较高，让人更容易将注意力放在会话上。人们利用这一空间进行简短的讨论和集中注意力。

Collaboration

"Block chairs" are used as seating for the coworking spaces. The light-weight stools can be easily stacked and moved, and can even be arranged into dividing walls, separating the office environment from the presentation and coworking spaces.

协作办公空间采用"积木凳"作为座椅。轻巧的小凳便于堆放和移动,甚至可以组成隔断墙,将办公区与展示兼协作空间隔开。

办公室的色彩搭配能给客户和其他访客留下强烈的印象,从而实现惊人的效果。这些效果作用在难以察觉的潜意识里,使他们对企业产生或好或坏的印象。

办公空间所运用的色彩应当随着自身的功能而变化。一推开门,人们就已经有了一个基本的印象和判断。正确选择办公色彩能保证企业传递出一个积极的形象。

在决定了想要传递的形象的前提下,办公色彩的搭配就变得十分简单了。一般来说,企业所在的产业越开放,色彩的选择越多、越富创造性。选择办公色彩搭配的最佳方式就是从公司的标志性色彩入手。

在本案中,鲜活的主色调是设计的核心。红、蓝、黄三个纯色分别标志着三个楼层,中间由色谱楼梯连接。墙壁上装饰着有关设计理念的手绘文字和图形。公司的吉祥物"艾普斯尔(Aposer)"是一个有两面形象的胖子。他的左面是无数的彩色糖果小人,另一面则呈现了内部器官,包括骨架、嘴、肠子、大脑、心脏等,这些器官由字母"Aposer"组成,代表了工作目标的统一和雄心。这个具有丰富意义的形象将二楼和三楼之间的楼梯平台连接了起来。

设计师:艾普斯托菲设计事务所　委托方:PIVIX 公司　竣工时间:2014 年　摄影:凯特斯里・旺万、希里查・伊昂维素特斯里　面积:176.41 平方米

APOS²
SPATIAL COLOURS REFLECT TEAM SPIRITS

Bangkok, Thailand

空间色彩体现群体的合作精神
艾普斯平方 / 泰国,曼谷

What to choose for the colours of an office can have a dramatic effect on clients and visitors. These effects are subliminal and instinctive, creating either a good or bad perception of the business. The colours used in an office should change with the type of business being conducted in the office. The image is being projected, and judgments being made, the moment a door is opened. The right choice for the office colours can ensure that the image is a positive one. Choosing the best office colour schemes is usually easy to do with some thought given to the image to create. In general, the less conservative of an industry a business is in, the more creative the office can be in terms of colours. A good way to start the process of choosing office colour schemes is to work with the company's signature hue.

In this case, vivid primary colours have been used as a core of design. Three pure colours red, blue and yellow have been marked in each level linked by a spectrum staircase and free hand Typographic and playful Pictogram which illustrate the design philosophy has been placed on the wall. There is a giant mascot named "Aposer", a figure of a fat man with two separate sides. On the left side, it is fulfilled with numerous colourful candies with different characters and faces while another side shows the internal organs such as the skeleton, mouth, intestine, brain and heart which are composed of the word "Aposer" representing unity and ambition as the goal of work. This humourous meaningful "Aposer" was figured to link the vertical space on a double-height wall at the landing between first and second floors.

Designer: Apostrophy's The Synthesis Server Co., Ltd.　Completion Date: 2014
Photography: Ketsiree Wongwan, Sirichai leangvisutsiri　Area: 176.41sqm

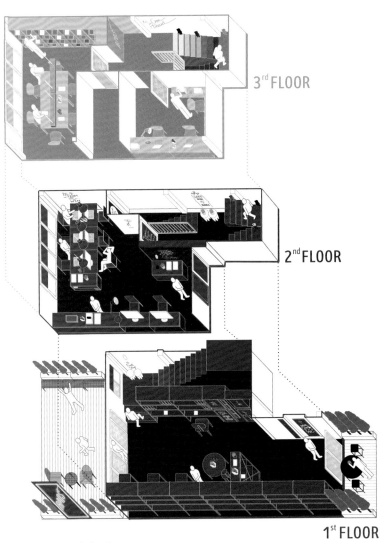

3rd FLOOR

2nd FLOOR

1st FLOOR

Concept Selection
概念选择

Collaboration

First Step – The retro neon sign has been placed on a hot red wall at the entrance. Then, entering to the ground floor the area has been designed to be a reception hall and café for whether officer or guest. There are chairs and long desk along the wall beneath a loft steel cabinet. Red is the most powerful warm tone colour so it has been used to energize, and stimulate a guest to be excited and staff to express their enthusiasm at the first step.

第一步——复古的霓虹灯被安装在入口火红的墙面上。一楼的入口区域被设计成接待厅兼咖啡厅,可供员工和访客使用。LOFG风格的钢制橱柜下方是一排座椅和长桌。红色是最强烈的暖色,它让访客一进门就感受到公司的活力,也让员工一进门就被激发出创作的热情。

Ground Floor Plan
1层平面图

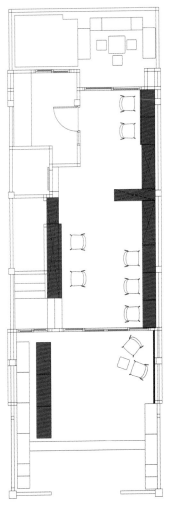

Collaboration

Keep Going – The first floor has been designed for working area called "Aposer Room", marked by blue, to make this space calm, tranquil and stable. It is an open plan office; all working desks have been placed facing each other to allow them to have some discuss. A huge typographic phase "There is no 'I' in TEAM but there is in WIN" has been placed on the wall dominantly to remind all staff to cut down their self-esteem, collaborate and have discussion to solve the problem or any conflict.

二楼是被称为"艾普斯尔室（Aposer Room）"的工作空间，以蓝色为主色调，显得冷静、平和、稳定。这是一个开放式办公空间，所有办公桌都面对面摆放，方便员工们随时讨论。墙壁上写着巨大的标语"团队不分你我，胜利属于我们（There is no "I" in TEAM but there is in WIN）"，鼓励员工放下自我，通力合作，解决难题。

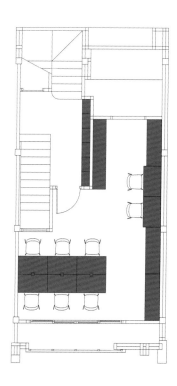

1st Floor Plan
2层平面图

Collaboration

To Continue – Whereas the second floor is the last level but is not the last of work, this floor has been realised for "Brain Storming Room" and executive room. It is marked by a bright yellow to light up the space and to spark creative thinking as well as the typographic phase about working development to inform both beginner staff and executive officers.

三楼是最后一层，被设计成重要的"头脑风暴室（Brain Storming Room）"和行政办公室。它以亮黄色点亮空间，激发员工们的创造性思维。墙壁上的创作标语更是无时无刻不在提醒着所有员工（无论是新人还是领导者）都要努力工作。

2nd Floor Plan
3 层平面图

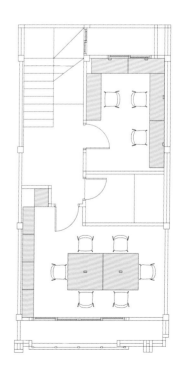

Collaboration

Axonometric View
工作台轴测图

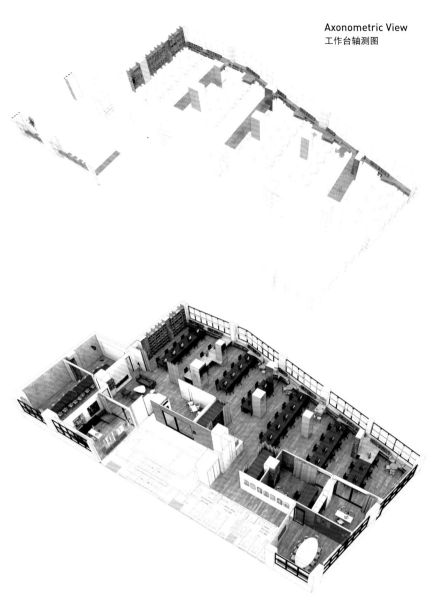

A creative workspace that "feels like home", designed to encourage new ideas and coordination between each team.

The design challenge was to create a generous studio-like workspace to reflect the company's creative culture, allowing cross-departmental staff to feel connected with each other and promote interactions within the company, as well as interactions with external partners or clients. The workspace successfully stems away from the old-fashioned corporately strict space, into a more relaxed atmosphere to encourage rigorous creativity.

In this case, the new workspace rejects the former offices' dated cellular cubicle design, to embrace a fresh open plan studio space, with long desk clusters forming the open plan areas. The designers also decide to use a continuous ribbon-like joinery system to encourage better communication and coordination that move around the floor space. While private offices and meeting rooms are located around the perimeter of the floor space, the design creates a variety of private, semi-private, and shared spaces.

Designer: Bean Buro   Client: Cheil Hong Kong   Completion Date: 2014

打造一个家一般的创意办公空间，这样的设计能激发新理念和团队间的合作。

设计所面临的挑战是打造一个宽敞的工作室式办公空间，以此来反映公司的创意文化，鼓励各部门之间的员工相互交流，促进公司内部以及与外部合伙人和客户之间的互动。办公空间成功地超脱于传统的严肃办公空间之外，营造一个轻松的氛围，能激发无限创意。

在本案中，全新的办公空间拒绝了老式格子间设计，呈现为清新开放的工作室布局，将长桌聚集在一起形成了开放的空间。设计师还利用连续的带状木制家具来鼓励人们视线更好的交流与合作。私人办公室和会议室设在楼面空间的外围区域。整个办公空间灵活多变，实现了私人空间、半私人空间和共享空间的组合。

设计师：宾·布罗   委托方：香港杰尔思行广告公司   竣工时间：2014年

CHEIL HONG KONG

A RIBBON-LIKE JOINERY SYSTEM HELPS INTERDEPARTMENTAL COLLABORATION

Hong Kong, China

带状的工作台有利于部门间的协同
香港杰尔思行广告公司 / 中国，香港

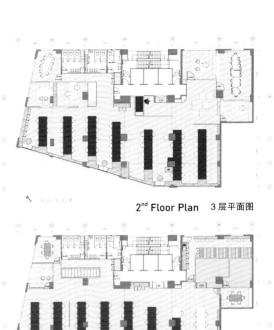

2nd Floor Plan　3层平面图

1st Floor Plan　2层平面图

Ground Floor Plan　1层平面图

The main concept was a continuous ribbon-like joinery system that flows and wraps around the perimeter of the floor space, linking all departments together with shared surfaces to encourage collaborations. The ribbon joinery system provides a variety of functions such as low storage shelves, in-wall seats for informal meetings and shared work surfaces that undulates between desktop and bar heights. Spawning from this ribbon around are the fingers of desk clusters, which mutates into full height bookshelves at one end. The overall effect is holistic and feels connected.

The material palette for the design feels collegiate, informal and relaxed. The ribbon-like joinery elements are made of recycled chipboard that is environmentally friendly. Seating upholstery and felt curtains provide acoustic and textural softness.

设计的主要理念是打造一个连续的带状木制家具系统。它将整个楼面空间从外圈包围起来，通过共享工作台将各个部门串联起来，促进了各部门之间的协作。带状家具系统具有多重功能，包含矮储物架、墙上休闲座椅、共享工作台等。办公桌像手指一样从带状家具伸出来，最终演化为墙边的落地书架。整体效果和谐统一，连贯自然。

设计所选用的主要材料给人以轻松自然的学院派风格。带状家具系统具有环保特征，由回收再利用的刨花板制成。座垫和毡帘显得柔和而温馨。

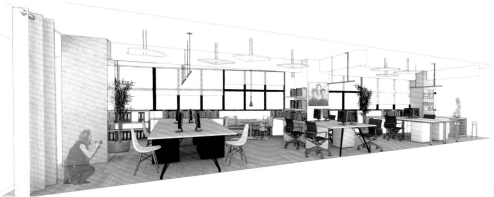

Rendering of the Ribbon
带状工作台效果图

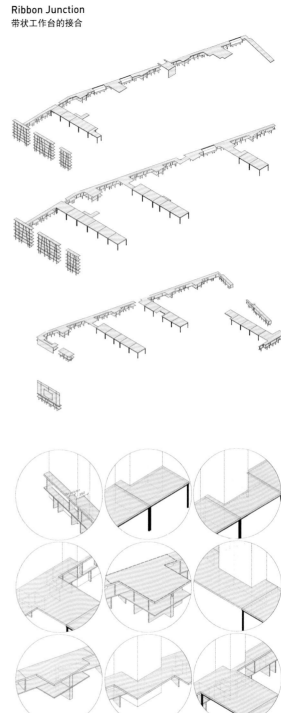

Ribbon Junction
带状工作台的接合

The design prioritises in giving generous spaces back to the staff, for them to feel like being at home and comforted to encourage coordination and creativity. Staff can gather for cosy idea discussions at various informal meeting places with other teams; a comfy lounge with a fireplace, casual in-wall café seatings, or tucked away in the custom-made diner booths. Staff are encouraged to personalise their environment and allow artworks to populate the walls and columns, subsequently creating a playful dialogue with the external renowned arts and craft district of Sheung Wan.

The lighting design is cosy and comfortable to complement the atmosphere that "feels like home". Gallery style spotlights highlight the artworks on the walls, while staff can enjoy meeting around the pendant lights in the ribbon joinery system.

设计将"为员工提供宽敞的空间"放在第一位，给他们家一般的舒适感觉，鼓励他们进行合作和创造设计。不同团队的员工们可以在各种各样的非正式会面场所进行舒服的探讨，可以是舒适的壁炉休息室、休闲的墙上咖啡座，也可以是定制的用餐间。员工们可以自主设计空间，在墙壁和柱子上添加艺术品，使办公室与以艺术和工艺闻名的上环地区实现灵活的互动。

灯光设计舒适自然，进一步突出了家的氛围。画廊风格的聚光灯突出了墙壁上的艺术品，而员工们则更喜欢在带状家具上方的吊灯下进行会面。

Collaboration

私人办公室和会议室设在楼层的外围。为了进行私密的对话，他们的墙壁是半透明的，营造出一种无等级的轻松氛围。落地玻璃墙上的半透明图形既是引导标识，又具有一定的遮挡作用。员工们还可以将正式或非正式讨论中获得的灵感写在黑板玻璃墙面上。

一组弯弯曲曲的会议桌设计精美，打破了传统企业会议室的严肃氛围。这些桌子有四种摆放方式，可适应不同的功能，既可以排成一排，用于大型会议；又可以组成小组，用于小规模会议。它的设计与不断变化的创意有着异曲同工之妙。

The private offices and meeting rooms are located around the perimeter of the floor space. For private conversations, their walls are translucent and create a non-hierachical atmosphere. Soft playful semi-transparent graphics are applied to the full-height glass walls to provide signage and control visual privacy. The wall surfaces made in black board glass allow ideas to be scribbled during formal and in-formal discussions.

A set of wiggly and curvy conference tables have been artfully designed to break down the strict atmosphere of a traditional corporate conference room. These tables can be playfully re-arranged into four configurations to adapt to different functions, from one long arrangement for large-scale meeting, to smaller clusters for small-scale meetings – a reminder of the constant state of transitions as the key to creativity.

LE CAMPUS

AN OPEN LAYOUT PROMOTES TEAM WORK

Paris, France

开放的布局方式促进团队工作

校园办公 / 法国、巴黎

In terms of how it improves teamwork and collaboration: Le Campus is a unique work space because it resides in a hotel, which is new. So while Le Campus is not technically a commercial office space, teamwork is encouraged by its open floor plan, without the confines of cubicles.

Virserius Studio's design idea was a departure from conventional thought about how people work, and their goal was that the space would embrace that same free and distinct thinking. The furniture used is modern and fun, but comfortable so that patrons feel welcome to share thoughts in a dynamic setting. Le Campus is a mostly open space, with the exception of its meeting rooms. These options were deliberate, a combination of formality and informality. Finally, the games and food in the Playground were incorporated into Le Campus to encourage connections, team-building, strategy and planning, all while having fun.

Design Company: Virserius Studio Designer: Therese and Regina Virserius Client: Hyatt Regency Charles de Gaulle Completion Date: 2014 Photography: Stéphane Michaux, Regina Virserius

在提升团队合作和协作方面,"校园办公"是一个位于酒店内部的新型办公空间。严格来讲,它并不是一个商业办公空间,它通过开放式布局来鼓励员工们实现团队合作。

Virserius 工作室的设计理念不同于关注人们如何工作的传统思维,他们的目标是让空间拥抱自由独特的想法。家具的设计现代而充满趣味,同时又十分舒适,让客户乐于在充满活力的环境中交换想法。这是一个几乎全部开放的空间,只有会议室是封闭的。设计经过了深思熟虑,巧妙地结合了正式与非正式的元素。"操场"上的游戏和美食在轻松愉快的氛围中促进了相互交流、团队建设以及策略和规划的生成。

设计公司:Virserius 工作室 设计师:特雷泽·维斯瑞尔斯、雷吉纳·维斯瑞尔斯 委托方:戴高乐凯悦酒店 竣工时间:2013 年 摄影:斯蒂芬娜·米修、雷吉纳·维斯瑞尔斯

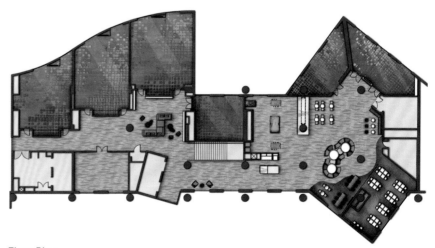

Floor Plan
平面图

Collaboration

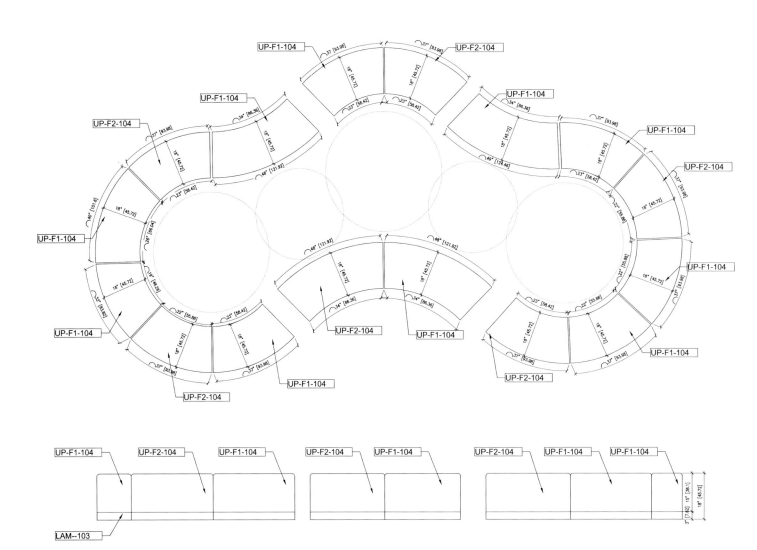

Custom Modular Sofa
自定义组合沙发

For some, the space may evoke the university days of yesteryear when one gathered with friends while studying for exams. For others, the bold colours and textures will offer a welcome respite from the more conventional office spaces.

"Le Campus was always supposed to be a very different working experience, something that will pleasantly surprise those with more traditional notions of a work space." The general manager of the Hyatt wanted something truly different for the hotel with the idea of going back to school, and approached V/S about designing a unique yet flexible space that could be used for work or social functions.

设计能让人产生重回校园的感觉，仿佛又与好友一起为了考试在图书馆中奋战。大胆的色彩和材质为人们提供了不同于传统办公空间的友好之感。

"'校园办公'项目给人以截然不同的办公体验，它会让持有传统办公观点的人们惊艳。"凯悦酒店的总经理希望为酒店打造一个重回校园的主题办公空间，因此他委托 Virserius 工作室打造一个独特而灵活的办公兼社交空间。

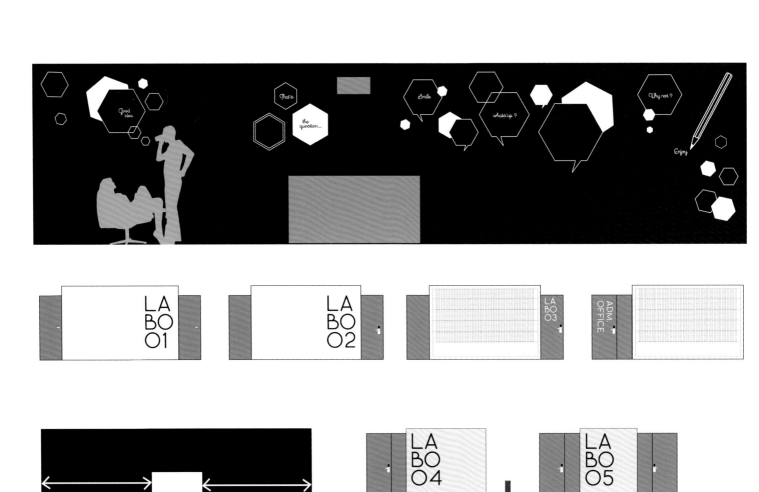

With inspiration from the American college campus, the designers created an experience that is interactive, cool, and playful. Graffiti artist Le Monstre was brought in to give the space an added edge. For others, the bold colours and textures will offer a welcome change from the more conventional office and meeting spaces. Meetings, events, and seminars are held in the "LABOs" (meeting rooms), the "Pique-Nique" (shared multifunctional space), and the Playground. Le Campus is truly a space that encourages and supports that highly creative and participatory experience enjoyed by the busy, on-the-move professionals who work, meet, and socialise during their time in this space.

设计师从美国大学校园中获得了灵感，打造了一个清爽有趣的互动空间。涂鸦艺术家蒙斯特为空间添加了额外的精彩。大胆的色彩和材质使其不同于传统的办公或会议空间。会议、活动和研讨会全部在"图书馆"（会议室）、"野餐区"（多功能共享空间）和"操场"进行。"校园办公"鼓励繁忙的差旅人士在这里获得创造性的互动体验，为他们提供了合适的办公、会面和社交空间。

Collaboration

LINKEDIN 605 MAUDE SUNNYVALE

AUXILIARY SPACES ENCOURAGE DIVERSIFIED TEAM INTERACTIONS

Sunnyvale, California, USA

辅助空间可促成团队互动的多样化
LinkedIn 领英公司森尼韦尔市莫德街605号办公楼／美国，加利福尼亚州，森尼韦尔

Collaboration is improved by providing a variety of environments, allowing individuals the freedom to choose locations based on the nature of the task at hand. There may be a need for privacy or a specific technology. The number of people involved varies from meeting to meeting. The diverse requirements of team interactions are supported in a collaborative office. Some examples of collaborative spaces in action include having break areas stocked with snacks, a tech lounge to provide assistance in keeping equipment updated and in good working order, meeting spaces that support a multitude of requirements, and giving a strong message that collaboration is more than OK. This is what makes a daily collaborative workplace.

In LinkedIn 605 Maude Sunnyvale, the design solution focused on the concepts of connection and transformation, both of core importance to LinkedIn company culture. The office space provides ample opportunities for employees to collaborate in a variety of ways; informal "chat rooms" for small groups adjacent to break rooms, relaxed living rooms for employees to put their feet up, formal conference rooms designed for comfortable acoustic and video conferencing interactions and a spacious game space with a barista bar for social gatherings.

Design Company: AP+I Design Designer: Carol Sandman, Larry Grondahl, Cailin McNulty, Jeff Baleix, Harland Patajo Client: LinkedIn Corporation Area: 14,400sqm Photography: John Sutton

1. Lobby Reception
2. Lobby Seating
3. Cafeteria / Servery
4. Cafeteria / Dining

1. 大厅接待处
2. 大厅座椅区
3. 餐厅／备餐室
4. 自助餐厅／餐厅

Ground Floor Plan
1层平面图

Collaboration

多样化环境的打造提升了协作价值，让个人可以根据手头的工作自由地选择办公地点。办公时人们可能有隐私或特殊的技术有所需求；各种会面活动的参与人数也是不定的。协作型办公室能满足团队互动的多样化需求。协作空间内设有储藏着零食的休息区；技术休息室能提供设备升级服务，使人享有良好的工作状态；会议空间能满足多方面的要求，传递强烈的协作信号。这才是优秀的日常协作办公空间。

LinkedIn 领英公司森尼韦尔市莫德街 605 号办公楼的设计方案主要集中在连通性和变形改造方面，这两点都是领英公司文化的核心要点。办公空间为员工们提供各种各样的合作机会：紧邻休息室的非正式"聊天室"可供小型团队举办会议研讨；休闲客厅让员工们尽情放松；正式会议室拥有舒适的视听会议互动设施；宽敞的游戏空间配有咖啡吧，适合社交聚会。

设计公司：AP+I 设计公司　设计师：卡罗尔·桑德曼，劳里·格伦戴尔，凯琳·麦克纳尔蒂，杰夫·巴莱克斯，哈兰·帕塔乔　委托方：LinkedIn 领英公司　面积：14,400 平方米　摄影：约翰·瑟顿

The ground floor is home to LinkedIn's Brick & Mortar Café with the fit, feel and food of a top rated restaurant, an innovative fitness centre, an airy lobby emphasizing the vertical connection between floors and abundant opportunities for innovation.

LinkedIn 领英公司的砖砌咖啡屋位于一楼，有一种高档餐厅的感觉。此外，一楼还设有创意健身中心和宽敞的大厅，后者突出了楼层间的垂直连接以及丰富的创新机遇。

The building interior is a vibrant space with emphasis on natural materials. LEED Gold objectives were supported with the use of cork, quartz and felt throughout the project. Wood flooring designates main circulation paths that create a sophisticated yet playful atmosphere.

建筑室内充满生机，突出了自然材料的运用。项目在各处使用了软木橡树皮、石英和毛毡，以实现 LEED 绿色建筑金奖认证。木地板标出了主要的交通动线，营造出精致而不失乐趣的氛围。

1. Game Room
2. Open Collaboration
3. Huddle Room

1. 游戏室
2. 开放协作空间
3. 小型会议室

1st Floor Plan
2 层平面图

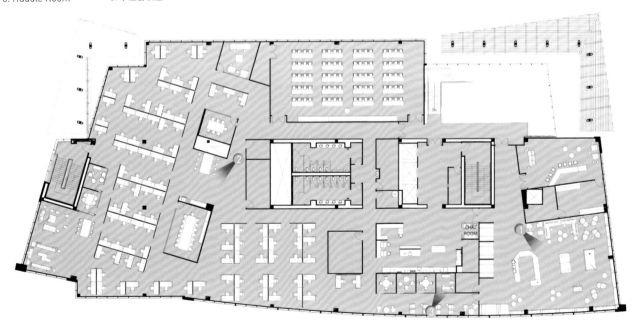

Collaboration

办公空间鼓励人们协同合作，并将此理念反映在各种类型的辅助区域中。员工们的工作台就像他们的"大本营"，他们还能随时进出各种辅助空间。当一个企业为员工提供了大量可用空间时，可替换的工作区域就显得至关重要，而流动性则是这种设计的关键所在。

Office space promotes the philosophy of collaboration, as is reflected in the various types of ancillary areas incorporated into the facility. Employees are given an assigned workstation to operate as a "home base" in addition to easy access to diverse support spaces. When businesses provide the staff with a broad range of spaces, it is clear that alternative work areas are important and mobility is part and parcel of making this work.

1. Open Office   1. 开放式办公区
2. Breakroom   2. 茶水间
3. Chat Room   3. 聊天室

Typical Floor Plan
标准层平面图

Collaboration

Selecting the right colour scheme makes a workspace dynamic and lively and improves staff's working emotion and efficiency, thus contributing to a company's development. The colours are also used as a means of orientation and colourful prints of open landscapes may create a relaxing atmosphere.

The project aims to connect flexible working and learning with brand awareness and to achieve maximum effects with limited methods. The colourful space represents the shining corporate identity. The interior design also integrates some enthusiastic elements which are set off by the colour printing of outside view. Varied working spaces are designed: work "lounges" in ICT department (with four different forms of work division), 1 person's focus rooms and also informal learning while sitting at the bar. The call centre seems an enclosed space with matching acoustics when you are seated, but as you stand up you enter a wide open space where you can have contact with your colleagues.

Designer: Liong Lie Architects Design Team: Liong Lie, Roeland de Jong, Michael Schuurman, Rajiv Sewtahal Project Management: Goed4U, Leidschendam Client: NTI Area: 2,200sqm Completion Date: 2014 Photography: Hannah Anthonysz

NTI OFFICE

THE FLEXIBLE USE OF COLOURS REFLECTS THE TEAM'S CORE VALUE

Leiden, The Netherlands

色彩的灵活运用可体现集体的核心价值

NTI 办公室 / 荷兰，莱顿

合适的色彩搭配能让办公空间变得更具活力、生气,使人们的工作情绪、效率得到改善,并且利于企业的发展。色彩还能起到导向作用,多彩的开放景观印花可以营造出一种轻松的氛围。

该项目的目的在于将灵活的工作与学习与品牌意识连接起来,用有限的方式得到最大功效。多彩的空间代表着NTI公司形象能够激情闪耀,内部设计也融合了热情的元素,这些颜色都比较合适并且用了外部景色的多彩印刷来烘托。有不同风格的学习区:工作区、一人学习室和吧台边的休闲式学习。当你坐下打电话的时候,电话中心像是一个封闭式空间,但是一站起来就可以直接与同事们联系。

设计师:Liong Lie建筑事务所　设计团队:良·列,罗伊兰德·德琼、迈克尔·舒尔曼、拉吉夫·修塔哈尔　项目管理:Goed4U公司　委托方:NTI公司　面积:2,200平方米　竣工时间:2014年　摄影:汉娜·安东尼斯

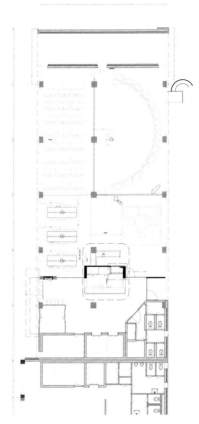

Ground Floor Plan
1层平面图

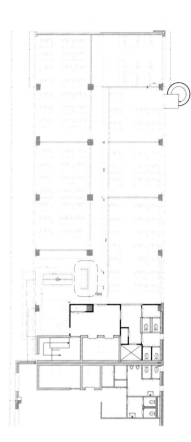

1st Floor Plan
2层平面图

The flexible working space could raise working efficiency and communication scintilla is bouncing. For NTI the team working is always the core value and the concept of the office design is to do all of the development and team collaboration in the open environment so that the NTI community can keep updated on what every member is doing and easily for them to make contributions to the company.

There's an open and welcoming atmosphere because of the use of open floor plans and bright colours.

灵活的办公空间能有效提高工作效率并促进交流。NTI公司一直以团队合作为核心价值，因此办公空间的设计让所有开发和团队合作都处在开放环境之中，让公司时刻了解员工们的手头工作，更便于他们对公司做出贡献。

开放式楼面和亮丽色彩的运用为公司营造出一种开放而友好的氛围。

2nd Floor Plan
3层平面图

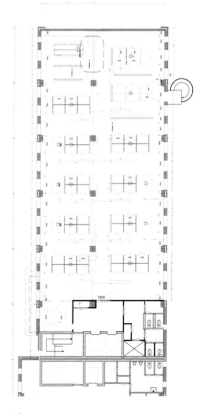

3rd Floor Plan
4层平面图

Collaboration

Flexible Working
灵活的工作方式

Forms of lersure to get re-energised
为了重新获得活力的休闲形式

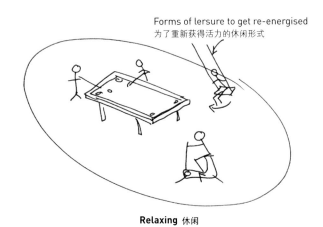

Relaxing 休闲

Colour bar
色柱

Capsules - with desk
 - as bench
封闭空间一带桌子
 一作为工作台

Concentrating 集中式工作区

Picnic table
茶歇桌

Carpet (zoning)
地毯 (分区)

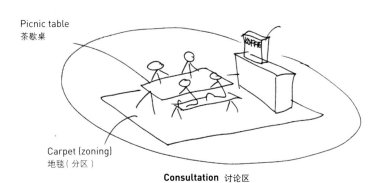

Consultation 讨论区

Dividers (new)
分隔板 (新)

Desks, re-use of dividers
可重复使用的分隔板

Colour bar
色柱

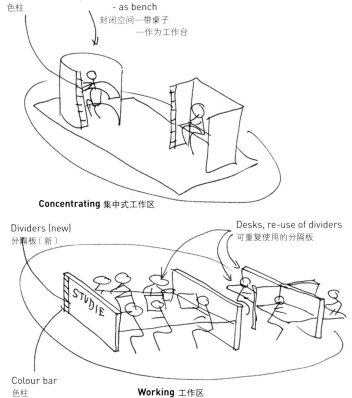

Working 工作区

The colourful corporate identity shines bright with enthusiasm and exactly that enthusiasm is incorporated in the design of the interior which could inspire the brainstorming and initiative of work. There's an open and welcoming atmosphere because of the use of open floor plans and bright colours.

Information collection and communication technology are no longer two separate stages overseen by different teams, but the work of one single leadership in the same space and time: ICT is finally central of the company.

色彩的运用是该项目最大的设计亮点，展现出激情闪耀的公司形象。这种激情能够激发员工的头脑风暴和工作热情。开放式楼面和亮丽色彩的运用让整个空间洋溢着一种开放而友好的氛围。

信息采集和通讯技术不再是两个容易被团队忽略的独立平台，而是在同一时间、同一空间下由单一领导所负责的工作，这使得信息通讯技术成为了整个公司的核心。

SPACES ARCHITECT'S @KA OFFICE

APPROPRIATE ZONING IN LINE WITH THE TEAM'S CREATION PROCESS

New Delhi, India

根据团队创意流程进行合理的分区

Ka 建筑事务所 / 印度，新德里

An office design was conceptualised to be a place where being leisure is also conductive for people to work in a creative environment, a workplace to enjoy. The zoning of spaces is justified keeping the main cabin with attached conference at the rear to maintain privacy as well as visually connecting it to front office.

In this case, the architect has maximised the spatial organisation of the place around the different design phases by choosing a central point, the green lounge, to make all the designers have a better cooperation. They have allocated a space to each phase of the creative process: display area; reception and waiting; design; experimentation; service area, to allow constant connections between the different design components: conceptual, technical and aesthetic. The creative process is not a linear mode of thinking. It changes course and veers of a tangent.

Principal Architect: Kapil Aggarwal Designer: Kapil Aggarwal, Pawan Sharma, Chander Kaushik, Karan Arora Site Supervision: Arvind Pal Singh Area: 139sqm Completion Date: 2013 Photography: Bharat Aggarwal

在概念化办公设计中,办公空间也是能够鼓励人们在创造性环境中努力工作的休闲空间,是人们享受工作的空间。空间的分区将会议室设在主办公区的后方,既能保证隐私,又能在视觉上与前方的办公室连接起来。

在本案中,建筑师在设计的各个阶段都力求实现空间组织的最大化。位于中央的绿色休闲区让所有的设计师能实现更好的协作。每个创意流程都配有独立的空间:展示区、接待和等候区、设计区、实验区、服务区等,保证了概念、技术、美学等不同设计元素之间的无缝衔接。创意流程不再仅仅是线性思维模式,它围绕着中心点不断变化方向。

主建筑师:卡皮尔·阿加瓦尔 设计团队:卡皮尔·阿加瓦尔、帕万·夏尔马、钱德尔·考希克、卡兰·阿罗拉 现场监管:阿尔温德·帕尔·辛格 建筑面积:139平方米 竣工时间:2013年 摄影:巴拉特·阿加瓦尔

Section
剖面图

Collaboration

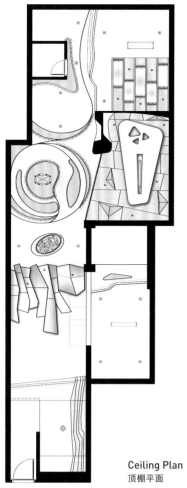

Ceiling Plan
顶棚平面

One moves down from front entrance highlighted by fixing glass roof penetrating ample light into the interior space. The reception table has an interesting form with backlit glass panel. A bookshelf displaying the firm's projects is placed opposite to it. A raised platform has an informal conference designed by fixing multiple dia. steel pipes with an abstract panel ceiling.

The ceiling plays an important role in the studio creating a visual transition between each department. As the ceiling near the reception made of multiple box panels continued to the ceiling in abstract form displays a journey of different projects and ideology of the firm. Similarly, elliptical ceiling over the reception has a hanging model inspired by Architect's Thesis Project being a focus in space.

入口的玻璃屋顶为室内空间提供了充足的自然采光。前台的造型十分有趣，配有背光玻璃板。展示着公司设计项目的书架正对着前台摆放。一个架高的平台上有一面由不同直径钢管支撑的休闲会议桌，上方的天花板设计也十分抽象。

在工作室中，天花板扮演了将各个部门连接起来的重要角色。靠近前台的天花板由多个箱式板材构成，上面展示着公司所设计的各种项目。类似的，前台上方的椭圆形天花板上也有一个建筑模型，构成了整个空间的焦点。

Collaboration

The flooring and walls at the front office is kept as cement finish to give emphasis on display panels.

The lower part has open workstations connected by cantilevered wooden steps, and the opening from front office looking towards the lower floor frames it. The lower part in contrast to the upper is designed in white tone.

前方办公区的地面和墙壁保留了水泥面,突出了展板的作用。

下层开放式工作区由悬臂式木台阶连接起来,前方办公区的开口正好与其相对。下层办公区与上层办公区形成了对比,采用了纯白色调。

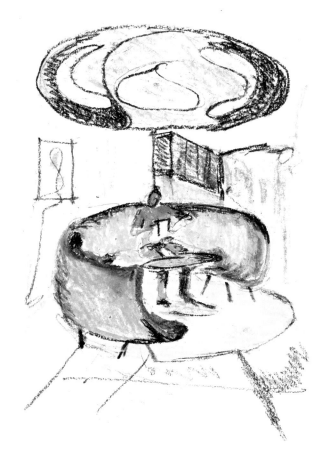

A green space with grass flooring and elliptical seating space is used as breakout space in the interior and used for reading books. The circular seating in green area is reflected on the ceiling in an abstract pattern continuing in the rear space.

Two workstations for senior architects are designed behind the seating.

人造草坪地面和椭圆形座椅所构成的绿色空间可用作休闲和阅读。环形座椅的设计被反映在天花板的造型上，延续了抽象图案的主题。

两个高级建筑师的办公台被设置在座椅的后方。

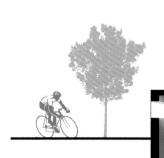

| Main cabin | Conference | Side entrance | Lower workspace | Meeting area | Section |
| 主区域 | 会议区 | 侧面入口 | 底层工作区 | 会面室 | 空间剖面 |

The most interesting part is an experiment with designing of the main cabin outer partition in a fluid form with veneer cladding continuing to the conference room ceiling. The partition is inclined at both the planes and takes an interesting form. The conference and cabin has a glass sliding folding partition which when pulled acts as individual space.

设计最有趣的一部分在于主办公区的流线造型外隔断，木皮包层一直延伸到了会议室的天花板。隔断略微倾斜，形成了有趣的弧线造型。会议室和办公区之间有一个玻璃拉门，可以实现独立的空间划分。

Collaboration

PART 2

EXCHANGE – FOR KNOWLEDGE'S EXCHANGE, SHARING AND INCREASE

第 2 章 交流型办公空间——有利于知识的交换、共享和增长

EXCHANGE AND THE DESIGN OF THE GLOBAL WORKSPACE

Exchange has been a formative aspect of workspace organisation and design since the inception of the traditional office. Technological development has transformed the way we interact socially and professionally and in turn changed the way modern business operates. The evolution of workplace design has been slower to catch up but the environment is changing, and information exchange is at the heart of this.

In living memory an international phone call pre-booked and physically connected through an exchange was a rare and costly event. Today according to the UN less than a lifetime later six of the world's seven billion people carry with them a device capable of making a call to the other side of the globe. Beyond the telephone, email and messaging have replaced the telegram, postal services and latterly the fax, and video conferencing to mobile devices enables face to face meetings to be held between disparate parties in the different time zones. Communication technology has shrunk the globe and accelerated the pace of life for better and for worse. We are now more connected than ever before constantly "jacked in" to a network, people inhabit the world oblivious of those sharing their immediate space preoccupied by the glowing screens of their smart phone. It is not unusual to see a group of people sitting together not talking or interacting with one another but each staring at the screen of their mobile device. Has this increased connectivity over long distance affected the way we interact face to face? How does this affect the workplace and how can the design of the workspace address this?

Technology has enabled remarkable changes to working practices; increased connectivity creates flexibility, facilitating opportunities for remote working and disparate collaboration. It is no longer necessary for all parts of the team to be located in the same space at the same time to do their jobs. Video conferencing and screen sharing enable co-workers to be thousands of miles apart and still able to work together, but does this effectively replace face to face creative collaboration? Most organisations whilst embracing the flexibility and global connectivity afforded by technology are still structured traditionally in nucleus workspaces gathering their teams under one roof and for many years as the communication devices and tools of the workplace evolved the office remained. The traditional office layouts remained unchanged, while the tools evolved – the

交流型办公空间设计

交流一直是办公空间组织和设计的一个重要组成方面。技术发展改变了我们人际互动和职业互动的方式，同时也改变了现代商业运作的方式。办公空间设计的进化正慢慢赶上，但是环境在变化，而信息交流方式是它的核心。

在记忆中，过去的国际电话还要提前预约并且得通过交换机连接，通话价格昂贵。现在，根据联合国的统计，全球70亿人口中的60亿都可通过通话设备与地球另一端的人随时通话。除了电话之外，电子邮件和短信也取代了电报、邮政服务和传真。视频会议让不同时区的双方可以进行面对面的交谈。通讯技术让地球变小，也加快了生活的脚步。我们现在前所未有地联系了起来，智能手机的屏幕在人们的眼前闪闪发光。如果一群人坐在一起，不说话，不交流，这并不奇怪，因为他们每人都盯着自己的移动设备。这种远距离的联系是否已经影响到了我们面对面的互动？这种现象又如何影响了我们的办公空间，而我们的办公空间设计又该如何应对呢？

技术让我们的办公发生了巨大的变化，更好的连通性创造了灵活性，实现了远程办公和远程协作。我们不再需要把团队中的每个人都聚集在一起就能展开工作。视频会议和屏幕共享让千里之外的合作人员可以一起工作，但是这能取代面对面的创意合作吗？在享受科技带来的灵活性和全球化联系的同时，大多数机构仍然选择通过传统的核心办公空间将他们的团队聚集在同一屋檐下，保留多年以来办公室所逐步形成的通讯设备和工具。传统办公布局保持不变，但是工具已经得到了进化——电脑取代了打字机，电子邮件取代了传真机，手机取代了固定电话。办公活动改变了，但是人们仍被禁锢在自己的办公桌旁。最近，移动技术实现了办公空间的灵活性，无线网络提供了更多的自由活动空间和选择，为跨领域合作、交互团队办公、协作、创新工作（现代商业的核心）提供了更多的机会。办公环境的设计将如何促进这种进步并利用现代技术呢？

今天，新一批的主要劳动力伴随着移动技术和社交媒体长大，

typewriter replaced by the computer, the fax by email, the telephone by the cell phone. Activities changed but still people remained shackled to their desks. More recently, however, mobile technology has enabled agility in the workplace; the absence of hard wired devices offers more freedom for movement and choice, which creates increased opportunity for crossover, interactions group working, collaboration, and creativity which is central to the modern business. How can the architecture of the workspace environment catalyse this and harness modern technology?

The emerging and soon to be majority workforce today have grown up with mobile technology and social media which is hard wired into the fabric of their communication repertoire. This literacy with multiple forms of communication extends into their professional lives and has blurred the boundaries between. Traditional modes of business communication have multiplied numerously requiring business to adapt to the plethora of ways in which their employees will communicate with each other, with customers, partners, vendors and suppliers. The modern workplace must embrace this and provide a carefully tuned environment for their team, which must combine well-managed and available technology with a well designed and collaborative workplace environment.

It is the role of the architect and designer to shape the workplace environment, to be carefully structured around a business' specific needs and the needs of their workforce in parallel with an understanding of their technological and digital requirements so the spaces function in concert with a company's tech and working practices. Mobile devices and wireless networking has freed the worker from their desk and have enabled us to be more nomadic in the workplace providing the opportunity for a variety of spatial conditions within a workspace to be created which can provide different environment for different forms of communication and work.

The designer must consider carefully the two fundamental ways people exchange in the workplace – face to face and across a network – analogue and digital. Networked exchange may be a phone call, a video conference, web chat, email, sms or numerous other messaging services. Face to face can be one to one, collaborative working, group meetings, training, presentation and conferences, but critically with relevance to the design of a workplace environment face to face is where the same physical space is inhabited. Video conferencing technology

这些东西已经成为了他们交流技能的一部分。这种多形式的交流模式已经进入了他们的职业生涯，并且模糊了工作与生活的界线。传统的商业交流模式要求商业在多方面适应员工与员工、顾客、合伙人、销售商、供应商之间交流方式。现代办公空间必须围绕这点展开并且为工作团队提供更适合的环境——必须将管理完善、合理实用的技术与设计良好的协作型办公空间结合起来。

建筑师和设计师的任务是围绕着公司业务的特殊需求和员工的技术和数字需求来打造办公环境，从而使得空间与公司的技术和工作实务实现统一。移动设备和无线网络把员工从办公桌前解放了出来，让我们在办公空间更具流动性，在办公空间内实现了各种各样的空间条件，适用于各种不同的交流和办公形式。

设计师必须认真考虑人们在办公空间中两种最基本的交流方式——面对面交流和网络交流。网络交流可以是电话、视频会议、网上聊天、电子邮件、短信或各种各样的信息服务。面对面交流可以是一对一、合作办公、团队会议、培训、展示演讲和会议，重点是面对面交流环境的设计与实际办公环境的设计是一致的。视频会议技术和远程呈现设备促进了数字化的面对面交流，但是数字技术在面部表情和人际互动细节方面的呈现仍需改进。

除了考虑个人与团队之间的交流之外，设计师还必须考虑到办公空间中交流模式对工作实务和他人工作效率的影响。例如，在一个活泼的开放式布局环境中，多个对话和背景噪声可能会影响需要集中精力工作的个人的工作效率。相反，一个活泼的环境也可以是激励因素，对某一项特定的工作产生积极的影响。移动技术让设计师可以打造各种各样的环境，分别适用于不同的任务和交流模式。

另一个必须考虑的问题是现代商务应当寻求在组织机构和各个层级之间发展一种"开放交流"和"可见性"的文化，办公空间的设计必须促进并鼓励这种文化的发展，打造一个能让人切实感受到开放交流的环境。

and telepresence devices improve digital face to face exchange but the nuances of facial expression and human interaction are yet to be successfully replaced by digital technology.

In addition to designing for exchange between individuals and groups, analogue and digital it is also important for the designer to consider the way that exchange in the workplace affects the working practice and efficiency of others, for example a lively open plan environment with multiple conversations and background noise can impact the productivity of an individual with a focused task to complete. Conversely a lively environment can be a motivating force and have a positive impact on productivity when suited to a particular task. The mobility of technology allows the designer to create a variety of environments suited to different tasks and modes of exchange.

Another consideration is the way in which a modern business should and does seek to develop a culture of "open communication" and "visibility" within the organisation and throughout the hierarchy and how the design of the workspace can facilitate and encourage this creating an environment where this sense of open communication can be tangibly perceived.

It is still common for a business to have embraced the technological revolution but their workspaces remain desk-based offices with several formal meeting rooms. We have found that whilst their staff communicate confidently across digital devices with one another and their broader network it is clear that the traditional form of sedentary allocated workstations discourages movement and interaction with individuals and between departments. The increase in digital communication has in turn in some cases reduced interaction both professionally and socially. A damning indictment of this is the entrance into the English language of the term "the water cooler moment" which refers to the sharing of casual gossip in the workplace centred on the only shared social space in the traditional office, the water cooler.

As a means to address this we seek to design workplaces which provide a variety of spaces which cater for and encourage varied working practices and modes of exchange from the single person private booth for focused working or sensitive conversation to the open collaborative forum or project space, to unshackle people from their desks and foster movement, crossover and creative collaborative

有很多公司虽然接受了技术革命，但是他们的办公空间仍然以办公桌为基础，配有若干正式的会议室。我们发现他们的员工习惯在私下里利用数字设备相互交流，很明显，传统的久坐不动的办公桌布局阻碍了个人以及部门之间的交流和互动。在一些情况下，数字交流的增加也相应减少了职业互动和社交互动。"茶水间时刻"正是这种状况的产物，指的是在传统办公室里，人们只有茶水间这一个共享的社交空间可分享社交八卦。

为了应对这一问题，我们所设计的办公设计需要提供能够满足各种工作实务需求和交流模式的空间。无论是适用于专注工作的单人工作间，合作团队共享的开放空间，还是把人们从办公桌前解放出来，促进交流、互动和创意工作的空间，它们必须有助于提升工作效率并具有多样性。虽然这些类型的空间设计能够促进专业互动和社交互动，但是偶遇空间和非正式交流空间的提供同样重要。大型办公空间中的节点岛能服务小型团体进行交流，例如，私厨岛或咖啡吧。中央咖啡吧或集会空间可以灵活地用于会面、娱乐、小型活动等形式，让员工暂时逃离办公室，共享轻松一刻。这些设计元素对打造一个友好而舒适的办公空间来说至关重要。在这样的空间里，办公室、咖啡厅、餐厅和家庭的界线将变得模糊。事实证明，这些空间对员工的身体健康和提高工作效率大有裨益。

要想成就优秀的办公空间，移动与流线设计也是至关重要的。它们能提高偶遇机会，促进团队以及部门之间的交流。传统办公楼的典型流线形式是开放式办公室配合长长的走道和垂直电梯。走道时连续移动的场所，没有停留之处，阻碍了人们的偶遇和对话；电梯是沉默的场所，对话在这里戛然而止。一名员工可能每天在电梯里看到同一个人三次，他们能认出彼此，但是从不交谈。设计师把移动作为一种积极的社交活动，在空间流线中创造出值得停留和偶遇的场所。这种人际交流对个人、团队乃至公司都有众多的益处。让我们用学校来进行类比，在英国的中学（学生年龄为11~18岁），每堂课都设在不同的教室，在课间，整个学校都在移动之中，这十分有利于促进自己核心群体之外的偶遇、对话和社交。这是一张更健康的社交和学习环境，能拓展我们的社交网络。

working, to provide spaces which enhance productivity and provide variety. These types of spaces are designed for both professional and social organised interaction, but the provision of places for chance encounter and casual exchange are vital. We use nodal local islands which serve smaller communities within the wider workplace which have the feel of a domestic kitchen island or coffee bar. In concert with these a well-designed light central café and gathering spaces is created, which can be used flexibly for meetings, entertaining, events and for the staff to escape the office relax and enjoy eating together. These design elements are vital to the creation of a welcoming and comfortable workplace where the boundaries between office, café, restaurant and home are blurred. These spaces are proven to enhance the sense of wellbeing and in turn the productivity of the workforce.

Also critical to the success of better workspaces is the choreography of movement and circulation to foster chance encounter and conversation between groups or departments who may not otherwise meet. The typical form of movement in a traditional office building is via long walkways through open plan offices and vertically in lifts. The walkway is a place for constant movement and offers nowhere to turn or stop; as a mode of circulation it eschews encounter and conversation, and in turn the elevator is a place of silence, where conversation stops. A worker may see the same person in the elevator three times a day, acknowledged through recognition, but they will never speak. We seek to utilise movement as a positive and socialising activity, choreographing a journey through the space whilst at the same time creating places stop and linger fostering encounter. This cross fertilisation across the workforce has numerous benefits to the individual, the team and the company. We often cite the analogy of a school: in the UK in secondary school for 11-18 year olds each class is taken in a different space, in between classes the whole school are on the move which creates encounter, conversation and socialising outside of your core group; this is a far healthier social and learning environment to broaden the analogue social network. Workspaces must learn from this and unshackle the workforce, remove departmental segregation and encourage movement and conversation.

The workspace designer must embrace and address all modes of exchange in the design of the modern workspace and vitally must communicate themselves with the workforce to address their specific needs and desires. At the outset of every project we undertake a thorough embedded research process with the client to understand the specifics of their team and their operation to ensure that their workspace is uniquely tailored to their needs and flexible enough to accommodate future evolution and change. This results in rigorously conceived and carefully crafted workspaces for the benefit of the workforce and the wider business.

Threefold Architects

办公空间也必须学习学校：解放员工，消除部门之间的隔阂，鼓励移动和对话。

在现代办公空间的设计中，设计师必须关注考虑到各种交流模式，他们必须亲自与员工交流，以便了解员工的特殊需求与渴望。在每个项目执行之初，我们都必须针对客户进行彻底的嵌入式调研，了解他们团队和运营的细节，以保证办公空间能够满足他们的需求并适用于未来的灵活改造。这样精心打造的办公空间将使所有员工乃至整个公司都受益。

Threefold 建筑事务所

ONE WORKPLACE HEADQUARTERS

THE EXPERIENTIAL SPACE ENCOURAGES IDEA EXCHANGES

Santa Clara, USA

体验功能的空间有助于思想的互动
第一车间公司总部 / 美国，圣克拉拉

The offices of new era need to embody their innovation and progression, especially for those offices with showroom experience. No longer a static showroom, the new working showroom need to demonstrate what is possible when great minds come together within the context of a multi-disciplinary design lab.

One Workplace is the largest furniture dealer in Northern California. The ambitious directive was to re-define the architectural standard not only of the company, but also of the showroom experience itself and create a bleeding-edge, world-class community that serves both employees and customers alike. One Workplace had already shifted the industry paradigm of sales and showrooms away from a transactional experience to one of collaboration and partnership.

The project consists of 35,000 square feet (32,252sqm) of office/showroom/workspace with an adjacent 180,000-square-foot (16,723sqm) warehouse (warehouse improvements were completed separately). In addition to the warehouse, the site also included an existing 10,000-square-foot (930sqm) stand-alone, mid-century office building layered with many years of dated tenant improvements. The project successfully connected this building and 25,000 square feet (2,322sqm) of warehouse into the re-imagined workplace. Design Blitz's design for the new façade and landscape improvements expand One Workplace's space into a multi-functional indoor-outdoor environment.

Designer: Design Blitz Area: 3,252sqm Completion Date: 2013 Photography: Bruce Damonte

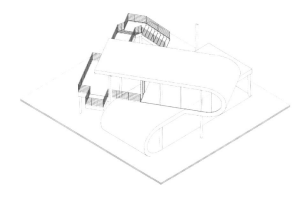

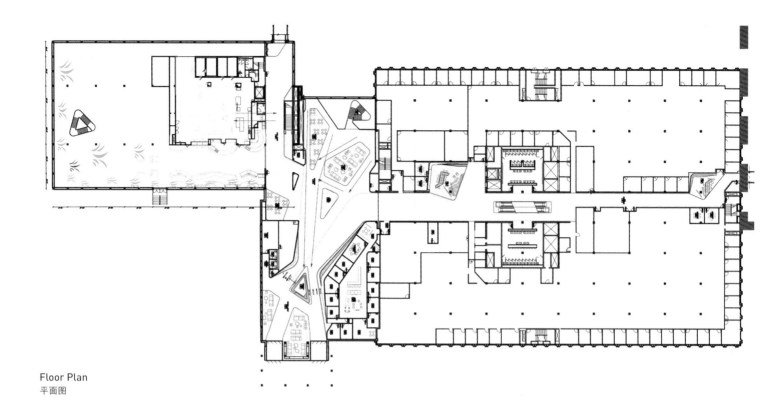

Floor Plan
平面图

新时代的办公空间必须体现创新和进步，特别是那些具有展示功能的办公空间。新的办公展示厅不再静止不动，而必须是一个多学科融合的设计实验室，把各种伟大的思想聚集起来，交流互动。

第一车间公司是美国北加州最大的家具经销商。项目的宏伟目标不仅是重新树立公司办公的建筑标准，而是重新树立展示厅设计的标准，为员工和顾客打造世界顶级的先锋社区化办公体验。目前，公司已经从事务型体验模式转化为合作型体验的产业标杆。

项目涉及 3,252 平方米的办公／展示／工作区和 16,723 平方米的仓储空间（仓库的改造独立完成）。除了仓库之外，项目场地还包括一个将近 930 平方米、建于 20 世纪中期的独立办公楼。项目成功地将建筑与仓库连接成一个全新的工作场所。Blitz 设计公司对外墙和景观的改造为公司提供了多功能的室内外环境。

设计师：Blitz 设计公司　　面积：3,252 平方米　　竣工时间：2013 年
摄影：布鲁斯·达蒙特

Sections of the "Boomerang"
"回旋镖"结构剖面图

1. Meeting　　1. 会议室
2. Discussion　2. 讨论室
3. Showroom　3. 展示区

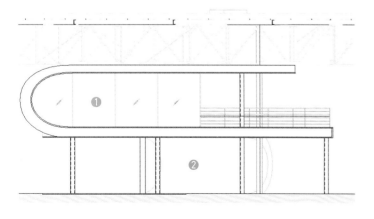
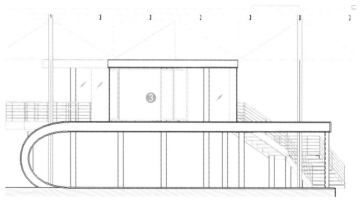

The boomerang further manifested itself in the creation of the two-storey stacked "boomerang" in the centre of the open office space. The elevated conference room and observation platform allows members of the One Workplace team to quickly survey the floor and show customers how a variety of systems solutions can intermix to create a unified, flexible and layered approach to workplace layout. The whole way through this workspace is the main street and the customers can feel the connection of this working community from here.

Design Blitz planned for longevity and flexibility by providing a raised floor system in the open office for easy future furniture reconfiguration as well as limiting colour and pattern to elements that are easily interchanged as trends change.

In learning more about how One Workplace engages with their clients Design Blitz determined that environmental context is key to effective sales. The community layout helps connect customers with the working staff. Customers will purchase an entire space, an environment rather than a single piece. It was crucial that the architecture supported the furniture and not the other way around. Individual spaces scattering around the main street were designed holistically to encourage an emotional connection by the customer.

在开放式办公区中央，一个回旋镖的结构形成了两层叠加的空间。架高的会议室和观景台让公司成员能快速检查楼面并为顾客展示如何将各种系统解决方案融入一个统一、灵活的分层工作空间。穿越这个空间的主通道让顾客们能感受到与工作社区的联系。

Blitz设计公司通过架高的楼面系统延长了开放式办公空间的使用寿命并拓展了它的灵活性，未来既能进行简单的家具配置，又限制了色彩和图案元素的运用，便于应对潮流的变化。

在了解了公司与客户的交流模式后，Blitz设计公司认为环境背景是决定实际销售额的关键。社区式布局有助于顾客与员工之间建立联系。顾客将购买整个空间、整个环境，而不是单一的家具。重点在于建筑辅助家具，而不是反过来。独立空间围绕着主通道分散开，采用整体设计，鼓励员工与顾客建立起情绪联系。

Upon entering the building you are immediately presented with the work-café. It is a public and social square of this workspace to meet and eat. Leading with this hospitality function ensures that customers and users encounter a warm and welcoming space. It was also during these initial conversations that Design Blitz began mapping both the client and user experiences through the space. The mapped experiences demonstrated that both customers and users would be sent out into the centre space to experiences a series of carefully planned touch points and then brought back to their starting point – much like the path taken by a boomerang.

走进大楼，首先映入眼帘的就是办公咖啡厅，这是一个可以会面和就餐的公共社交广场。这种待客功能让顾客和用户能感受到温馨友好的氛围。在与委托人的初步交流中，Blitz 设计公司就已经决定将这里作为一系列体验的起点。根据规划的体验路线，顾客和用户将进入中央空间体验一系列精心规划的触点，最后又回起点，就像回旋镖的来去路线一样。

Exchange

Exchange

In addition to being an innovative design, the project demonstrates strong metrics for the economics of efficiency. One Workplace moved from a 45,000-square-foot (4,180sqm) space into the new 35,000-square-foot (3,252sqm) space while increasing staff from 101 to 165. The increased efficiency was achieved by reduction in workstation foot print and a move by the majority of the sales team to a mobile work flow where workers do not have a dedicated workstation. Mobile workers store their belongings at a centralised location and work either at a shared workstation or in the soft seating of the work-café or alternate work areas. One Workplace is walking the walk when it comes to modern work typologies.

除了创新设计之外，项目还具有极高的经济效益。公司从4,180平方米的空间搬进了3,252平方米的空间，而员工人数则从101人增至165人。工作台占地面积的缩减以及销售团队的移动办公（大多数销售人员都没有固定的办公桌）都是效率提升的关键。移动办公人员可以将他们的物品储存在集中的场所，然后在共享工作台、办公咖啡厅的沙发上或其他办公区域工作。第一车间公司正全面迈进现代办公模式。

OFFICE MINDMATTERS

A HIGHLY FLEXIBLE WORKSPACE TAPS THE TEAM'S POTENTIALS

Hamburg, Germany

高度灵活的空间有利于团队潜力的发挥
Mindmatters 公司办公室 / 德国，汉堡

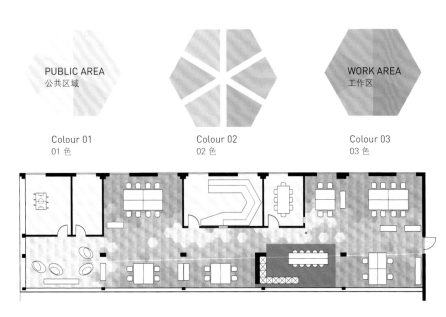

A relaxing workplace can benefit your business by enhancing job satisfaction and reducing stress. Such working offices for the fast-growing company need an open space balanced between areas for interactivity and for retreat. This helps employees to contribute more insights and strive to increase their responsibilities, maximising the potential of a small group of staff.

PARAT has created a new office space for Mindmatters, a software developer in Hamburg. Two central conference rooms divide the floor plan and create niches which house working zones. The design is playful and graphic without looking childish. As such, it is built on soft, natural tones punctuated by yellow.

The new office has been designed for a team that spends much time in the office. This is why the main emphasis was put on the social and meeting spaces – natural materials and cosy seating furniture helped create a relaxing atmosphere for better communication. The wooden arena with cosy seating furniture is the main feature of the office space as well as the whole interior.

Designer: PARAT   Area: 360sqm   Completion Date: 2014
Photography: Andreas Meichsner

一个轻松的氛围能提升员工对工作的满意度、缓解压力，从而使企业受益。快速成长型公司需要开放的空间来实现互动性与休闲性的平衡，这有助于员工贡献出自己的见解和能力，从而提升他们的责任感，实现团队潜力的最大化。

PARAT 为汉堡的一家软件开发公司 Mindmatters 设计了全新的办公空间。两个中央会议室将楼面隔开，形成了壁龛式的办公区。设计趣味十足，活泼多变，又不会显得幼稚，以柔和自然的色彩为基调，辅以黄色进行点缀。

新的办公室是专为大部分时间都在办公室工作的团队所设计的，因此重点放在社交和会面空间上——天然材料和舒适的座椅家具有助于打造轻松的交流氛围。被舒适座椅所环绕的木台办公空间乃至整个室内是设计的主要特色。

设计师：PARAT 设计公司   面积：360平方米   竣工时间：2014年   摄影：安德里亚斯·密切纳斯

Exchange

The patterned floor is the key element of the design: it is an extensive customised mosaic of carpet tiles with a range of colours and surfaces. The dark grey marks the workspaces, brown is used for areas intended for movement, while light carpet is used to define the public zones such as the library and conference rooms.

The heart of the office is a long table with a kitchen unit. The table is divided in half, allowing a floor lamp to be integrated and seating to be doubled. Another meeting space is the Mindmatters conference room with a wooden arena for upwards of 20 people. The construction enters into dialogue with the characteristics of the room.

图案化地面是设计的关键元素：特别定制的块式地毯具有丰富的色彩和纹理。深灰色标志着工作区，棕色用于通道区域，而浅色地毯则代表着会议室、图书室等公共区。

办公室的中央是一个配有厨房设施的长桌。长桌被一分为二，整合了落地灯和座椅。另一个会议空间是可以容纳20人的木台会议室。整体设计施工与房间的特色形成了对话。

Exchange

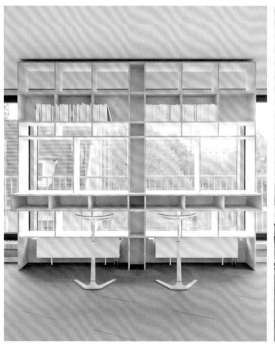

Each project team has a mobile shelf with writable doors and an inbuilt 60" screen for reviews: this item acts both as storage space and as a room divider at once. The staff doesn't have their own table. This causal working help inspire their brainstorm. The back area of the office provides further opportunities for team work or communication in beanbag chairs. A floor-to-ceiling shelf offers work stations with standing aids and frames the view out to the neighbourhood.

每个项目团队都有一个移动书架，配有可书写的门和60英寸显示屏。书架既是储藏空间，又是房间隔断。员工没有自己的办公桌，这种轻松的氛围有助于激发他们的头脑风暴。办公室的后部适合团队工作或私人交流，配有懒人沙发。落地书架与工作台是一体的，透过窗口可以看到外面的风景。

办公室希望保留建筑的工业特色，突出它物理性质和复古工业气质。员工们需要一些更开放的空间和设施来实现更好的交流：开放的办公桌、舒适而私密的会面区、吧台高的工作台和小会议室、大厨房、集体午餐区（兼作展示区）、大会议室、自行车库和非正式办公区。

重比特实业是一个全新的会员制办公空间，专为制作云开发产品的创业公司所设计。"重比特项目"通过特邀展示、公共活动、教育、建议和开发者合作来辅助、培养并促进这个相对较新的计算机产业分支的创新、方案解决和企业发展。根据以上的功能要求，项目在原有的建筑外壳内嵌入了一系列的设计改造。

设计师：IwamotoScott 建筑事务所　竣工时间：2012年

HEAVYBIT INDUSTRIES

DIVERSIFIED FACILITIES PROMOTE STAFF'S COMMUNICATIONS

San Francisco, USA

变化多样的设施促进员工思想的交流

重比特实业 / 美国，旧金山

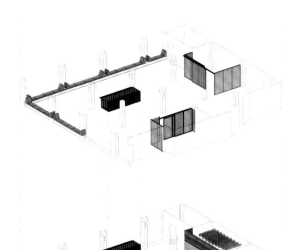

Diagram of Dividers
隔断示意图

The work office aims to retain the industrial character of the building and to emphasize its physical and vintage industrial qualities. The working staff requires more open space and facilities for the better communication: an open array of desks, comfortable and intimate meeting areas, bar-height workspace and conference rooms, a large kitchen, collective dining area for daily catered lunches that could double for speaker presentations, conference room, bike storage, and informal work areas.

Heavybit Industries is a new members-only workspace designed for early stage companies making cloud developer products. The Heavybit programme curates, fosters and promotes innovation, solution-finding, and business development in this relatively new branch of computing through invited presentations, public events, education, advising, and inter-developer collaboration. Given the project brief, the programme is addressed through a series of designed interventions inserted into the existing shell.

Designer: IwamotoScott Architecture　Completion Date: 2012

Exchange

The largest of the interventions is a multi-functioning platform at ground level with a new stair leading to the first floor. The platform, which helps create a more communicational environment, constructed as a "solid" laminated plywood object houses the reception desk located opposite the main entry, bar-height work counter that doubles as seating on the raised platform side, speaker stage facing the dining area, pass-through ramp, and U-shaped lounge seating. Located to spatially subdivide the ground floor and create different areas on each side, while keeping it visually open, the platform also serves as the first landing for the new stair to the first floor. Surrounding the platform on one side is a communal dining and meeting space, with Y-shaped tables that work for communal meals with causal communication and to seat an audience for events.

一楼的多功能平台与新建的楼梯相连，直通二楼。平台采用胶合板构造，形成了一个更适合交流的环境。平台具有一些设施：前台与正门相对，吧台高的工作台也可以作为座椅，演讲台正对就餐区，同时还设有坡道、U形休闲座等。平台对一楼空间进行了划分，在两侧形成了不同的区域，同时又保证了空间的视觉开放性。此外，它还是通往二楼楼梯的第一个楼梯平台。平台的一侧是公共就餐和会面空间，配有Y形的休闲桌，可用于就餐、休闲交流或在活动中充当观众座椅。

The main stair itself is suspended from a series of ½" × 3" (1.3×7.6cm) steel fins welded to steel channels framing the new cutout in the first floor, which also become the stair risers. The affect of the stair is at once heavy and light. From the rear it appears as a series of steel plates, from the side it almost disappears. The hexagonal brake-formed perforated steel used between the fins affords this transparency and also refers to the company logo of the hexagon.

主楼梯悬挂在一系列1.3cm × 7.6cm的钢片上,而钢片则焊接在二楼的钢槽上,同时也是楼梯踢板。楼梯的效果在轻盈中又不失厚重。从底部看,它是一系列钢板,从侧面看,它近乎隐形。呈六边形镂空的钢网在钢片之间形成了半透明的保护围栏,同时也与公司的六边形LOGO相呼应。

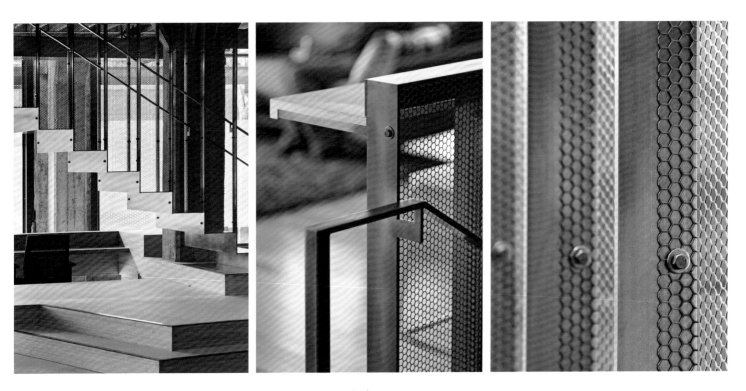

Exchange

在二楼,钢片向上延伸成为护栏和台面高的会议吧台。二楼和三楼的办公空间被规划成开放式办公环境,设有书桌面板和若干个会议室。会议室被设计成滑动墙的形式,滑动墙采用Polygal面板、钢框和金属螺丝构造,同时还配有特别定制的钢制谷仓门五金和轨道。非正式办公和会面空间在二楼沿着窗口展开,配有连续的软垫窗台座;在三楼则以独立的房间呈现,采用电缆和绳索分隔出来。与楼梯相似,"绳索室"也给人亦轻亦重的感觉;它由打结的工业棉绳构造而成,有一种悬空的感觉。

On the first floor, the fins extend upward to become a guardrail and counter-height meeting bar on one side. The first and second floor workspaces are planned as an open working environment with desk pods and several conference rooms on each floor. The conference rooms are made to appear as a set of sliding walls constructed of Polygal over steel frame and painted exposed metal studs. These also incorporate the custom steel barn door hardware and track. Informal work and meeting space is made along the window on the first floor with a new continuous felt-upholstered window seat, and on the second floor by a freestanding suspended room constructed of electrical conduit and rope. Similar to the stair, the "rope room" is at once heavy and light; it is made from thick, cotton industrial rope with heavy knots, yet it also floats in the room in both plan and section.

The ground floor conference room and bike room are defined by a reclaimed wood and glass wall, and the kitchen combines dark grey stained plywood cabinetry and industrial fittings for pull hardware. The steel of the stairs reappears at a detail level framing the wall opening and custom barn-door track, and in the kitchen as a large, folded steel island countertop.

一楼会议室和自行车库采用回收木材和玻璃墙隔开，而厨房则综合了深灰色斑点胶合板和工业化五金件。楼梯所使用的钢材在细节中反复出现，变身为窗框、拉门轨道以及厨房里的钢铁台面。

这个大型项目中有三项装置采用了先设计后建造的原则进行制作，它们分别是："绳索室"、厨房天花板上的"六角蜂巢钢"吊灯和二楼会议室的"六角蜂巢布料"吊顶装置。两个"六角蜂巢"装置都采用了六边形图案，以不同的方式与公司的 LOGO 相呼应。六角蜂巢钢吊灯由薄薄的波形黑钢和爱迪生灯泡构成，体现了老建筑的精神，而六角蜂巢布料则是一个轻质的拉伸吊顶结构。后者由普通廉价的无纺布网制成，弹性布料从各个方向被拉紧，在房间里形成了漫射灯光效果。与楼梯和平台一样，这些装置都试图发掘普通材料的潜力，赋予它们全新的意义。

Three installations within the larger project were commissioned on a design-build basis. These include the "Rope Room", "HexCell Steel" light over the kitchen island, and "HexCell Fabric", a ceiling light diffuser in the first floor conference room. Both "HexCell" installations use a hexagonal plan pattern that recall the Heavybit logo, but in different ways. The HexCell Steel light is made from thin, contoured, blackened brake-formed steel with Edison lightbulbs in the spirit of the existing building, while the HexCell Fabric is a lightweight tensile ceiling structure. Made of ordinary and inexpensive non-woven mesh, the ceiling was designed using a physics modeler so that the flexible fabric is pulled taut equally in all directions, creating a geometrically precise but diffuse light effect in the room. In each case, as with the stair and platform, these installations attempt to defy the predictable qualities of the ordinary materials of which they are made.

TRIBAL DDB AMSTERDAM

AN OPEN WORK ENVIRONMENT HELPS INDIVIDUALS EXPRESS THEIR CHARACTERISTICS

Amsterdam, The Netherlands

开放的空间氛围激励员工个体特色的表达
阿姆斯特丹 Tribal DDB 公司 / 荷兰，阿姆斯特丹

An innovative working environment where creative interaction and communication happened is supported and to achieve as much workplaces as possible in a new structure with flexible desks and a large open space. All of this while maintaining a work environment that stimulates long office hours and concentrated work.

Tribal DDB Amsterdam is a highly ranked digital marketing agency and part of DDB international, worldwide one of the largest advertising offices. i29 interior architects designed their new offices for about 80 people.

The design had to reflect an identity that is friendly and playful but also professional and serious. The contradictions within these questions asked for choices that allow great flexibility in the design.

Situated in a building where some structural parts could not be changed it was a challenge to integrate these elements in the design and become an addition to the whole. i29 searched for solutions to various problems which could be addressed by one grand gesture.

Designer: i29 interior architects Area: 650sqm Completion Date: 2014

这个创意办公环境能促进创新互动和交流，在新建筑结构内通过灵活办公桌和大型开放空间的设置实现了办公空间的最大化。整体设计形成了一个能让人长时间集中精力工作的办公环境。

阿姆斯特丹 Tribal DDB 公司是一家行业领军的数字化营销广告公司，隶属于全球最大的广告公司之——恒美广告公司。i29 室内建筑事务所为他们拥有 80 多人的办公室进行了设计。

设计必须反映出公司的形象——友好幽默，专业严肃。这些观点的矛盾要求设计必须具有极大的灵活性。

由于办公室所在的建筑有一些结构是无法改变的，因此对室内设计提出了一定的挑战。i29 针对各种问题——找出解决方案，最终实现了完整的设计。

设计师：i29 室内建筑事务所 面积：650 平方米 竣工时间：2014 年

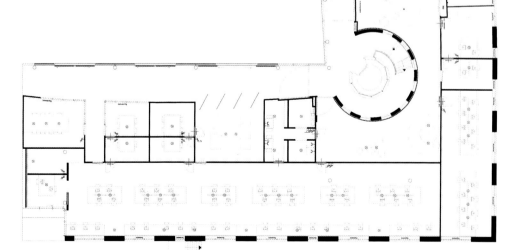

Ground Floor Plan
一层平面图

Exchange

The office space situates in the space without other structural parts and the designers decide not to change the whole simple and focused atmosphere. The main design concept is the openness of working and communication. The company believes that openness builds efficiency. Separate workshops may push the one choice to the limit. So the designers decide to use open spatial pattern with simple and linear table to give energy in this space without fail into chaos. But more importantly it leaves the workers with a charismatic environment. The white long table on the other side helps the workers leap into communication easily and at the same time would not interrupt the other teams.

Compared with the traditional cellar office space, this new office could help strong individuals to express one or few characteristics. This whole working atmosphere encourages the workers from different disciplines to work collectively and the arced half-open meeting/lounge could promote better exchanges and cooperation.

办公空间所在的位置没有其他的建筑结构,而设计师也决定保留这种简单而专注的氛围。主要设计概念是开放的工作和交流空间。公司坚信开放能产生效率。独立的工作间可能会限制人的选择。因此,设计师决定采用开放式空间布局,以简单的直线办公桌来为空间注入活力。更重要的是,它能为员工带来富有魅力的环境。另一侧的白色的长桌能让员工轻松交流,同时又不会打扰其他团队。

与传统的格子间办公室相比,这个新办公室能帮助强烈的个体表达自身的特色。整个办公氛围能鼓励不同领域的员工共同办公,而弧形的半开放式会议室/休息室则能促进他们的交流和协作。

In order to integrate to the structural parts, at first a material which could be an alternative to the ceiling system, but also to cover and integrate structural parts like a big round staircase. Besides that, acoustics became a very important item, as the open spaces for stimulating creative interaction and optimal usage of space was required.

This led the designers to the use of fabrics. It is playful, and can make a powerful image on a conceptual level, it is perfect for absorbing sound and therefore it creates privacy in open spaces. The designers used it to cover scars of demolition in an effective way. There is probably no other material which can be used on floors, ceiling, walls and to create pieces of furniture and lampshades then felt. It's also durable, acoustic, fireproof and environmentally friendly, which doesn't mean it was easy to make all of these items in one material!

为了与建筑结构相融合，设计师试图寻找一种能够覆盖并整合结构元素的天花板系统。除此之外，隔音也是重要的考量之一，因为公司需要一个能刺激创意互动、优化空间利用的环境。

最终设计师选择了布料。它具有趣味性，并且能形成强有力的形象，同时还能吸音、在开放空间内打造私密环境。设计师用布料有效地将拆除过程中所形成的"疤痕"覆盖起来。可能再没有一种材料能同时应用在地面、天花板、墙壁、家具和灯罩上了。布料经久耐用、隔音、防火、环保，但是整体制作过程却并不简单。

BRIGHTLANDS CHEMELOT, BUILDING 24

VARIOUS SOCIAL SPACES SATISFY STAFF'S INTERACTIVE REQUIREMENTS

Limburg, The Netherlands

多样的社交空间满足员工互动的需求
切美洛特化工园 24 号楼 / 荷兰，林伯格

Designers choose to use some facilities to provide the welcoming feel of a home in some large industrial buildings: it would be a sitting area with sofa, a kitchen with dining tables and a loft with meeting rooms. All of them form a pleasant resting and meeting place.

Broekbakema has since 2006 been working on completion of the master plan for Chemelot Campus, a breeding ground for research talent from DSM and other top companies in Sittard-Geleen. The Limburg-based campus is regarded as a source of technical innovation in The Netherlands and a driver of economic growth in the region. In partnership with Studio Niels™, Broekbakema has designed a relaxing working environment for these research professionals a central large industrial facilities building to serve as a campus living room that stimulates encounters and cross-fertilisation of ideas.

Designer: Studio Niels™ and icw BroekBakema Client: Brightlands, Chemelot Campus Completion Date: 2014
Photography: Serge Technau

设计师利用一些设施在大型工业化建筑中营造出家庭的温馨感:可以是配有沙发的休息区、带有餐桌的厨房,也可以是阁楼会议室。这所有的一切形成了愉快的休闲和会面空间。

自2006年以来,Broekbakema一直致力于切美洛特化工园的总体规划。切美洛特化工园为研究机构和顶尖的公司提供了产业摇篮。这个位于林伯格的园区是荷兰技术创新的圆圈,也是整个区域经济发展的驱动器。Broekbakema与Niels™工作室共同为这些研究人员设计了一个轻松的办公环境,这座位于园区中央的大型工业设施建筑是整个园区的活动中心,能鼓励人们相互交流,促进行业发展。

设计师:Niels™工作室、icw BroekBakema设计公司 委托方:切美洛特化工园 竣工时间:2014年 摄影:泽格·泰克诺

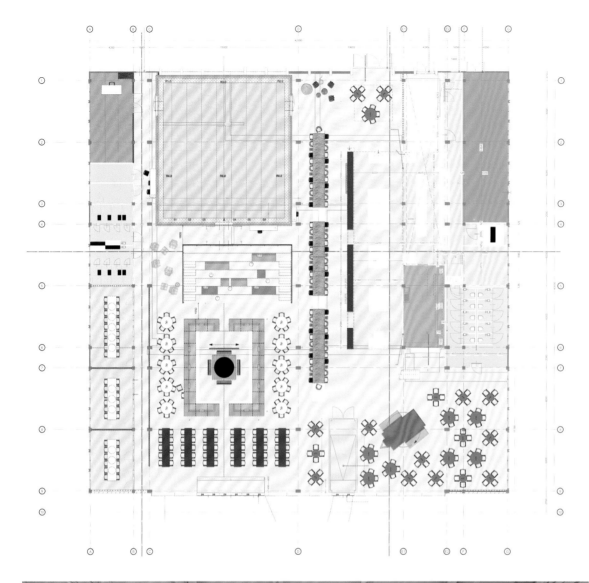

Floor Plan
平面图

Transformation into a Transparent Meeting Place

The choice fell on redesigning part of the existing Building 24, as this will enable the desired facilities to be brought together in a central location. Opting for the transformation of an existing structure rather than new-build is a choice for sustainability in optima forma. Although the characteristic building was originally to be demolished, Broekbakema was able to demonstrate the architectural quality and potential practical value of the double-arched hall. This has resulted in the preservation of the hall with only a few interventions needed to transform it into a vibrant meeting place. By introducing a large glass wall – the most important change – the north façade has become fully transparent.

通透社交空间的改造

项目对24号楼的部分空间进行了重新设计，将一些设施聚集起来。选择对原有空间的改造而不是新建一个空间，是出于可持续发展的考虑。尽管园区的特色建筑已经被拆除，Broekbakema 力求在本项目中证明双拱大厅的建筑质量和潜在实用价值。设计师保留了大厅结构，仅做了少量改动，将其改造成活跃的社交空间。作为最重要的改动，大面积玻璃墙的应用让北立面变得完全通透起来。

Flexible Interior that Invites Interaction

Different functions are brought together in the restyled Building 24. Broekbakema and Studio Niels™ have designed and positioned sport, meeting and restaurant facilities in a way that encourages visitors to interact. For the client, this way of using space can be seen as an experiment, as the intended cross-fertilisation gives the campus an innovative boost. Inherent in this design is a learning ability: the interior of the building is highly flexible, so it can easily be adapted to user experience and changing requirements.

Studio Niels™ created a layout of an ordinary house with Living Room (corporate yellow sofa), Living Kitchen, Dining Room (large dining tables), Staircase Room, Winter Garden with a flower bomb (the kiosk). Hospitality & Working are not separate in this concept; they are seamlessly integrated.

The idea is all about sitting together in the

state of mind people are in; for example I want to meet on an informal base and therefore I choose to sit on the sofa and hang out and chat for example. You like working on a kitchen table and meet the rest of the family so you choose the large dining tables, where the history of the company is reflected in the design of table in the kitchen tile prints. All objects are designed to meet, whether you are alone or you are together.

A huge success factor for creative workspaces is the coffee bar especially when there's a barista at work; the designers placed the LaKiosk as a flower bomb in the space to add lots of colour and created a fresh and happy feeling for the working people.

The designers created an open kitchen layout where you pick your plate out of the kitchen cabinet and walk through the kitchen and crab your food and drinks. They placed fresh herbs in the cabinet to give this more boost, and suggested using the cabinet for personal art and book space.

The concentration workplaces offer more privacy by the design and also by the position in the layout, the coloured Breakout Rooms in the sides of the building are more private rooms and these rooms offer Hardware tools for presentations.

The designers created the large icons so they transformed a functional element also to a decorative element.

鼓励互动的灵活室内空间

不同的功能区汇聚在重新设计的24号楼内。Broekbakema设计公司和Niels™工作室设计并布置了体育、会议、餐厅设施，鼓励人们进行互动。对委托人来讲，这种空间使用方式可以看作一种试验，人们的交流互动将为园区注入创新活力。设计的内核是学习能力：建筑高度灵活，随时可以根据用户体验和变化的需求进行改造。

Niels™工作室打造了一个普通的住宅布局，包含客厅（企业代表色黄色的沙发）、厨房、餐厅（大餐桌）、楼梯间、冬日花园（设有花朵咖啡亭）。服务设施和工作空间并没有分离，而是紧密地融合在一起。

设计概念让人们可以随心选择自己所需的空间。例如，我想要进行一次非正式会面，那么我就选择沙发，来回走动着聊天；你想在小厨房工作，与家里人会面，那么就选择大餐桌（餐桌上的打印材料介绍了公司的历史）。所有物品都为了会面所设计，适合各种规模的会谈。

在创意办公空间的设计中，一个极大的成功因素就是咖啡吧，特别是有专业咖啡调配师的咖啡吧。设计师把咖啡吧设计成花朵的海洋，用丰富的色彩给工作人员带来了清新愉悦之感。

在开放式厨房中,你可以从橱柜拿出餐盘,穿过厨房,随意挑选美食和饮料。设计师在橱柜中放置了新鲜的香草来注入活力,并建议人们在橱柜中放置个人艺术品和书籍。

集中办公区的设计提供了更多私密感,多彩的休息区设在建筑的两侧,因此更加私密,这些房间还配有用于展示的硬件设施。

设计师所打造的大图标既是引导标识,又具有装饰性。

GOOGLE TOKYO

DECORATIONS WITH VARIOUS COLOURS CREATE A RELAXING COMMUNICATION ATMOSPHERE

Tokyo, Japan

通过不同色彩的装饰营造轻松的沟通氛围

谷歌东京 / 日本，东京

To help increase communication between staff and make them feel comfortable and prevent visitors from becoming lost, more and more designers decide to define various zones with specific colours and patterns and also give each a distinct character. This develops a way to expand company's facilities that isn't repetitive or boring and which also assists wayfinding.

This ambitious interior project is located in the Roppongi Hills tower in central Tokyo. Each zone was assigned a specific colour, the colours being modulated through different tones. This creates a "necklace" of differently coloured meeting rooms, each with a specific name and character, strung around the building's large central core.

Designer: Klein Dytham architecture   Completion Date: 2014

为了促进员工之间的交流,使他们感到舒适并防止访客迷路,越来越多的设计师开始选择用特定的色彩和图案来定义不同的空间,以赋予每个空间独立的个性。这种设计不仅能扩展公司的设施,使其不会重复无聊,还有助于寻路。

这个宏大的室内项目位于东京市中心的六本木之丘大厦。每个功能区都被赋予特定的色彩,并且以不同的色调呈现出来。色彩各异的会客室各具特色,配有不同的名字,它们围绕着大厦中央核心空间分布,形成了一条"项链"。

设计师:KDa 建筑事务所 竣工时间:2014 年

On one of the floors, KDa defined the circulation route around the meeting rooms with the perforated concrete block walls common in Tokyo's winding residential lanes. In the city these block walls often provide glimpses into lush gardens, and KDa used them here to allow views into enticing spaces beyond the walls. Each of these "pocket parks" has a huge wall graphic of brightly coloured plants and can be used for gatherings and informal meetings. These "pocket parks" create a relaxing and flexible communication environment which can inspire cosy discussion.

在其中的一个楼层,KDa 建筑事务所围绕着会客室打造了一个环形路径,设计选择了东京民居小巷常见的镂空混凝土墙。在城市中,我们通常可透过这些墙壁看到里面郁郁葱葱的花园。KDa 在这里用它们实现透镜效果。每个"迷你公园"都配有巨幅壁画,上面绘有色彩亮丽的植物,可用于聚会和非正式会面。"迷你公园"的设计营造了一种轻松、灵活的交流环境,鼓励人们进行愉快的讨论。

KDa also placed landmarks at key positions to help staff and visitors identify their location and navigate around the floor. KDa have provided mini-kitchens where staff can grab snacks and drinks, each space decorated a different colour. After having designed kitchens themed by Google colours – blue, yellow, red, green – on the lower floors, KDa then looked to create something even more memorable: a bright blue "hairy kitchen" clad in the giant brushes used in automatic carwashes. All the relaxing space with different bright colours can help the staff have a brainstorming anywhere.

A set of spaces on another floor was themed after a sento, the traditional neighbourhood bathhouses now fast disappearing from Japan's cities. Passing through a traditional noren curtain, leads to space instantly recognisable as a "wash area", complete white ceramic tiles, wooden stools, and computer screens cunningly configured where mirrors would be expected. This leads on to a spacious "soaking bath" area – actually a presentation and training room – which like classic sento features a huge mural of Mount Fuji specially created for Google by one of Japan's last living mural painters. This space is also used for external events, with the "wash area" becoming a reception space for drinks and catering. Nearby, a group of meeting rooms have a matsuri (traditional neighbourhood festival) theme. Here, red and orange wallpaper picks up patterns from the yukata robes and happi coats worn at festivals, wall graphics show photos of festival scenes, and sake and beer crates both act as impromptu seating and create a relaxed party atmosphere.

KDa还在主要位置设置了地标，帮助员工和访客实现定位和楼层内的导航。KDa为员工提供了迷你厨房，让他们随意享用零食和饮料，每个空间都采用不同的色彩进行装饰，分别配以谷歌的标志性色彩——蓝、黄、红、绿。在下面的楼层，KDa试图打造一些能带给人更深印象的空间：亮蓝色的"发丝厨房"整个被全自动洗车房所使用的巨型刷子所包围。色彩亮丽的轻松空间有助于员工们随时随地进行头脑风暴。

另一个楼层的的一系列空间被设计成了"钱汤"（日本一种传统的公共澡堂，在当今的城市中正迅速地消亡）主题。穿过传统的门帘，映入眼帘的是显而易见的"洗浴区"：白色瓷砖、木凳、镜子。电脑屏幕被巧妙地设置在镜子的位置，隐藏起来。再往里走就是"浴池区"——实际上是展示和培训室。就像传统的"钱汤"一样，墙壁上是大幅的富士山壁画，壁画是全日本仅存的几位壁画画师之一专门为谷歌创作的。"洗浴区"变成了接待空间，配有餐饮服务。旁边的会客室采用"乡宴"主题。红、橙色的壁纸采用日式浴衣上的图案进行装饰，墙上的挂画展示着节庆的场景，日本米酒箱和啤酒箱都是临时的座椅，营造出一种轻松的派对氛围。

Floor Plan
平面图

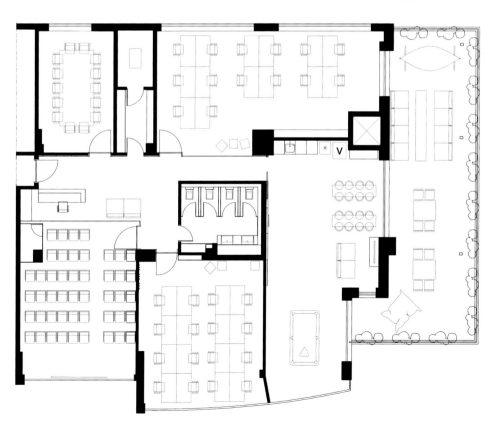

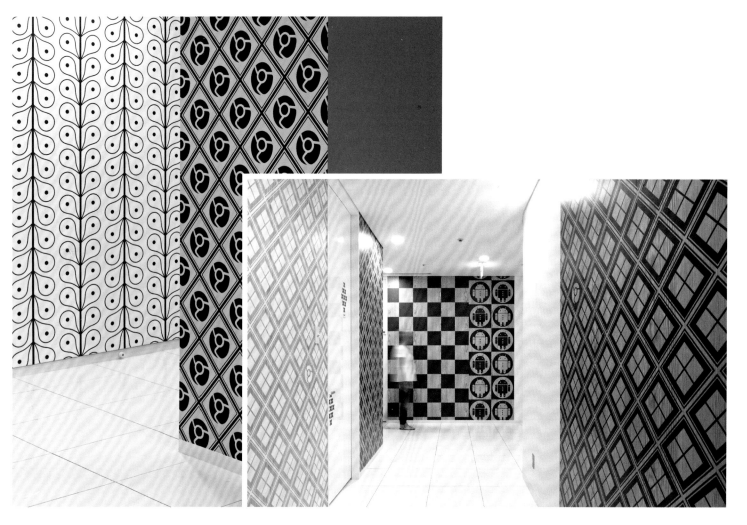

Google request that each of their national offices around the world reflects the unique culture of its location. KDa's design for the earlier phases of the project had taken cues from the graphics of traditional Japanese fabrics and contemporary anime, but then Google requested an even more vivid evocation of Japanese culture. Looking to communicate the Japanese context without resorting to cliché, KDa incorporated surprising elements such as a full-scale yatai (mobile food stall) and a digital koi pond that greets people at one of the entrances – responding to hidden sensors, carp projected onto the floor move towards those who enter the space as if expecting to be fed.

For previous sections of the interior, KDa created brightly coloured wallpaper patterns cleverly derived from refigured Google icons such as the Google Android and Google Map pin. For the new spaces, KDa developed a set of muted, timber-coloured wall graphics whose tone varies from light to dark wood. Subtly evoking Japan's traditional timber architecture, the patterns occasionally incorporate cunningly hidden icons.

谷歌要求全球的办公室都要反映该地区的文化特色。KDa对项目的早期设计从日本传统纺织品的图案和现代动漫中获得了灵感，但是谷歌后来要求体现一种更为鲜活的日本文化。为了深入体现日本文化，又不落入俗套，KDa引入了一些惊喜的元素，例如等比例的路边摊，门口迎接宾客的数码锦鲤池等。在锦鲤池的设计中，投影器受隐藏的传感器触发，将锦鲤投射在来人身前的地面上，就像是祈求投喂一样。

对于先前设计的各个部门，KDa采用从谷歌机器人、谷歌地图别针等图标中演化出色彩明亮的壁纸进行装饰。在新空间的设计中，KDa开发了一套柔和的木色墙壁图案，呈现为或深或浅的木色。这些图案既隐喻了日本传统的木建筑，又巧妙地隐藏着各式图标。

KOIL – KASHIWA-NO-HA OPEN INNOVATION LAB

EFFECTIVE CONNECTIONS BETWEEN AREAS FACILITATE INTERDISCIPLINARY COMMUNICATIONS

Kashiwa-shi, Japan

各区域的有效连通促进跨领域沟通
柏市开放创新实验室 / 日本，柏市

The innovative workspace needs to include various functions for smooth communication crossing over fields, such as having a meal, manufacturing, making presentations and relaxation. Users may choose their place from various locations within the complex freely and work as sharing the spaces and facilities with other users, which allows users to contact with others diversely. The office will become a place like a miniature of urban city, where various activities and events occur simultaneously.

KOIL (Kashiwa-no-ha Open Innovation Lab) is an innovation centre intended to support start-ups of entrepreneurs, promote enterprise developments and stimulate economic activities in Japan. The staff would fully use skills and know-how to produce innovative products and services, which is facilitated and realised by the system with investors' supports. In contrast with offices during the 20th century which were made up based on uniform flexibilities for managers, this space represents the flexibilities of communication for each worker.

Designer: NARUSE INOKUMA Architects, Co., Ltd. Total Floor Area: 2,576sqm (6F)
Completion Date: 2014 Photography: Masao Nishikawa

Exchange

创意办公空间需要包含各种有助于跨领域交流的功能区，让人们在就餐、生产、展示、休闲的时候尽情交流。用户可以自由地选择空间，与其他用户共享空间和设施，实现各种方式的交流互动。办公室将变成一个微缩的城市，同时进行各种活动。

柏市开放创新实验室是一个致力于帮助创业者的创新中心，其目标在于促进日本的企业发展、刺激经济活动。企业员工们利用技能和专业知识来制作创意产品和提供创新服务，而投资系统则为他们提供帮助和支持。与20世纪以经理人为基础的灵活办公空间相比，这一空间为所有员工都提供了灵活的交流空间。

设计师：NARUSE INOKUMA 建筑事务所　总面积：2,576平方米（6楼）　竣工时间：2014年　摄影：西川政尾

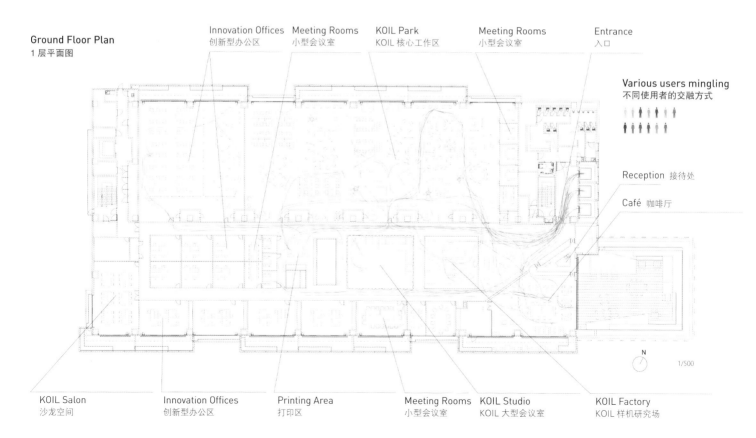

For the place allowing flexible activities, the diverse space has been created with areas that have different applications and intersect within the central public zone and with various ceiling heights, colour temperature of the lights, and finishes of interior designed to match the areas' communication functions. It seems to increase restrictions to limit the characters of spaces and usages; however, the layout functioning organically altogether enables the space to meet all kinds of working style of all users.

在灵活空间的设计中，各个区域都有不同的应用配置，它们与中央公共区域相互连接，以各种各样的天花板高度、灯光色温、室内装饰来配合各个区域的交流功能。这似乎限制了各个空间和功能的特点，但是从根本上融合起来的功能布局让空间能配合所有用户的各种办公风格。

About the visual elements, the unfinished atmosphere is emphasised in the design aiming to make people's activities themselves to be the charm of the space. The ceiling of share area has exposed duct pipes reflecting the lights. The other elements are simply finished using the base materials such as wood wool cement boards, fibre-reinforced cement boards, and plaster board levelled with putty covered with clear paint. Through such details, the design intends to avoid restricted meanings or a particular style imaged from the final finish.

在视觉元素的设计中，设计师力求用一种未完工的氛围来突出人们的活动，使他们成为空间的魅力所在。在共享区的天花板上，裸露在外的管道反射着灯光。其他元素也都采用了最基本的材料进行装饰，例如水泥木丝板、纤维水泥板、涂清漆以油灰找平的石膏板等。这些细节设计旨在避免形成限制感和固定的装修模式。

Exchange

MOZILLA VANCOUVER OFFICE RENOVATION

THE BREAK OF CLOSED OFFICE AND THE FREE EXCHANGES BETWEEN EMPLOYEES

Vancouver, Canada

封闭办公区域的打破与员工的无障碍交流
摩斯拉公司温哥华办公室 / 加拿大,温哥华

The Internet company is a highly collaborative not for profit organisation with a mission to promote openness, innovation and opportunity on the web. In such a renovative office there should be no receptionist, no corner offices and a focus on spaces that reinforced collaboration between individuals and teams. What was once a maze of strangely shaped offices in traditional business and dark hallways in traditional office is now a warm and inviting space much better suited to Internet company's friendly open culture.

Mozilla leased a prime 580-square-metre space in the heart of Vancouver's emerging downtown tech neigbourhood. The space had originally been designed to accommodate typical office tenants with enclosed offices along the perimeter and open plan work areas in the centre of the floor plate. The plan reinforced standard office hierarchies and limited interaction and collaboration amongst staff. The design process leads to a solution where meeting rooms and support spaces were internalised and the open plan workstations were distributed at the exterior of the floor plate.

Designer: Hughes Condon Marler Architects   Completion Date: 2014   Photography: Martin Tessler   Construction Cost: $679,000 CAD

这家互联网公司是一家协作型非盈利组织，其目标是促进互联网的开放、创新与发展机遇。在这样的办公室里，没有接待员，也没有角落办公室，空间的重点在于加强个人和团队之间的合作。曾经布局砌块的传统办公格局和昏暗的走廊已不复存在，取而代之的是温馨、迷人的空间，更适合互联网公司友好、开放的企业文化。

摩斯拉公司在温哥华市中心的新兴技术区内租用了一个580平方米的顶级办公空间。这一空间属于典型的传统办公布局：封闭的办公室沿着外围走廊展开，楼面中央是开放式办公区。这种空间规划突出了标准办公层级，限制了员工之间的互动和协作。在翻新改造中，设计方案实现了会议室和辅助空间的内化，并且将开放式办公台分散在楼面的外围。

设计师：HCM建筑事务所　竣工时间：2014年　摄影：马丁·特斯勒
建造成本：679,000加元

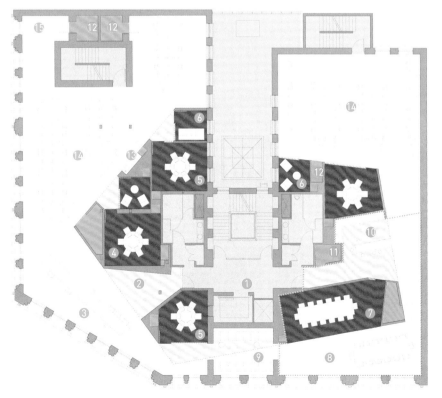

Floor Plan
平面图

- Base building spaces　　基础空间
- Meeting rooms　　会议室
- Storage space　　存储空间
- Extent of overhead volume　　空闲区域
- Boundary of meet up space　　空间边界

1. Public lobby
2. Semi-public lobby
3. Standing / sitting co-work table
4. Medium meeting / black box room
5. Medium meeting room
6. Small meeting room
7. Large meeting room - enclosed
8. Large meeting room - not enclosed
9. Casual meeting / workspace
10. Kitchen
11. Server / communications room
12. Storage
13. Coffee bar
14. Workstations and co-work spaces
15. Fitness area

1. 公共大厅
2. 半公共大厅
3. 站立 / 合作工作台
4. 多媒体会议室 / 暗室
5. 多媒体会议室
6. 小型会议室
7. 大型会议室 – 封闭式
8. 大型会议室 – 开放式
9. 临时会议室 / 工作区
10. 厨房
11. 通信机房
12. 储物间
13. 咖啡吧
14. 工作站 / 合作工作区
15. 健身区

This strategy freed up the perimeter zone for casual meeting and working spaces, giving unobstructed access to communication and interaction between everyone. Chalkboard walls, pin-up spaces and formal and informal meeting spaces are arranged such that the collaborative culture of the client is reinforced. Interactive work spaces were equally accessible to everyone no matter what their role might be.

全新的布局解放了外围区域,使其成为休闲会面和办公空间,实现了人与人之间的无障碍交流和互动。黑板墙、便利贴空间以及正式和非正式会面空间的布局突出了公司的协作文化。互动办公空间让各种身份的工作人员都能平等进出。

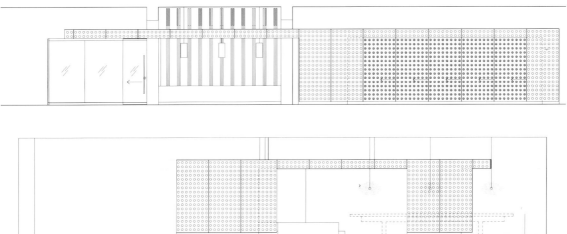

The design placed meeting rooms on the interior of the floor, rather than the exterior - providing access to large swaths of natural light. The red causal meeting space achieves architecturally what Mozilla stands for as a company and its people. It creates a kind of meeting space that would make the invisible communication visible.

设计将会议室设在楼面的内层，使其能够获得细长的自然光线。红色的休闲会面空间在建筑形式代表了摩斯拉公司和它的员工。这种会面空间的设计似乎能让无形的交流变得可以触碰。

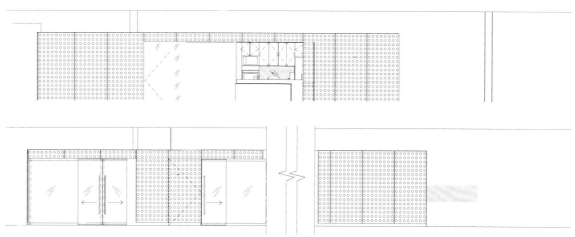

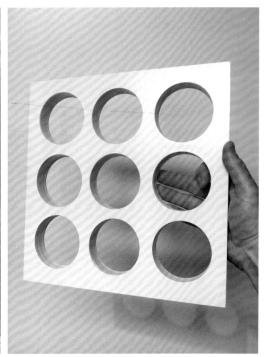

Exchange

PART 3

COMMUNITY – GREAT INCLUSIVITY TO SPACES COMBINING WORK AND SOCIAL ACTIVITIES

第 3 章 社区型办公空间——对工作与社交相结合的场所有极大包容性

COMMUNITY BY DESIGN

1. The Relativeness of Physical Space

Mobile technology plays a key role in the hyper-connected, hyper-engaged, highly informed and mobilised society we live in. Laptops, mobile phones (and all the gadgets you can name), have disassociated the modern work dynamics from the physical space that was meant to shelter them. The office, as it has been conceived for the last decades is now obsolete and useless.

The new generations of workers, digitalised and hyper connected are free of any physical or time limitations; hence they can work whenever and wherever they want, with no need of even leaving their home. It is not needed to file documents physically anymore. You don't have to be with someone to have a meeting. It is not essential to deliver by hand any report, and nothing happens if I am not in my workplace, because I don't need to be there anymore.

There is nothing new or avant-garde in this statement; it is a phenomenon that has been here for a while, but it is important to realise that this change of paradigm makes necessary a profound reflection about the workplaces typologies and how it must adapt to these new work dynamics.

Historically, when faced with the problem of designing a workplace, one of the main drivers has been capacity of space. But the problem has changed, and thus a new question must be asked.

From "How many workers can be fitted by square metre, in order to maximise production?", now the question to be answered when designing a workplace must be "How do I improve the work experience of the inhabitants to optimise their performance?" And this new question is a game changer. First, because when evaluating a workplace, quantity is replaced by quality. Second, because the focus of design is changed from the construction of a built environment to the materialisation of a shared experience. And a shared experience is what better determines and identifies the creation and permanence of any community. It is a dramatic shift from production-centred design into community-centred design.

2. Workplaces Redefined by Unresolved Work Dynamics

As traditional offices have failed to respond to these new standards, people have left these workplaces and have searched for what Ray Oldenburg has come to call third places, which he defines as the social surroundings separate from the two usual social environments of home and the workplace.

"Oldenburg calls one's 'first place' the home and those that one lives with. The

社区型办公空间设计

1. 物理空间的相对关系

在我们所在的高度连通、极度繁忙、信息爆炸的移动社会中，移动技术扮演着重要的角色。笔记本电脑、移动电话等一系列移动设备让现代工作动态与物理空间分离开来。现在，办公室已经变得过时，显得可有可无。

新时代的工作者，在数据和移动设备的武装下，不受任何物理限制或时间限制，可以根据需求随时随地地工作，甚至在家也能工作。因此，无需再进行物理归档，交谈也不必当面进行，更不用亲手递交报告。即使不在办公室也不会影响我的工作，因为我已经不再需要办公场所。

以上所描述的状况并非新鲜事，这种现象已经持续了一段时间。重要的是，这种改变让我们认识到有必要对当前的办公空间进行反思——如何使它们应对全新的办公活动。

传统来说，在面临办公空间的设计时，最主要的驱动力之一就是空间的容量。但是现在我们所面临的问题已经产生了变化，必须应对全新的问题。

在办公空间的设计中，问题从"为了实现生产率的最大化，在有限的面积内要安排多少工作人员？"变成了"如何提升使用者的办公体验才能提升他们的办公效率？"

这个新问题为我们带来了巨大的转折。首先，在评价办公空间时，质量取代了数量。其次，设计的重点由建筑空间的建造转移到了共享体验的实体化。共享体验更能决定整个团队的创造表现。办公空间的设计已经从以生产为中心的设计变成了以社区为中心的设计。

2. 未处理的办公活动重新定义了办公空间

既然传统的办公室已经无法应对这些新要求，人们就离开了这些办公空间，去寻找社会学家雷·奥登伯格所提到的"第三空间"，即脱离于家庭和工作场所两种常见社交环境的社交场所。

"奥登伯格将人的'第一空间'称为家庭，是人生活的地方。'第二空间'为工作空间，是人们度过一生中大多数时间的空间。'第三空间'是社区生活的'锚点'，能促进更广泛、

'second place' is the workplace – where people may actually spend most of their time. Third places, then, are 'anchors' of community life and facilitate and foster broader, more creative interaction. All societies already have informal meeting places; what is new in modern times is the intentionality of seeking them out as vital to current societal needs."[1]

Let us agree on something: It is not comfortable to work in a Starbucks. There are few sockets, tables are generally unstable and round shaped (worst geometry for laptop activity), and the risk of dropping your coffee over your computer or documents is particularly high. It is not a quiet and serene place either; on the contrary, there are lots of visual and acoustic distractions, privacy is minimum and going to the restrooms raises the dilemma: should I carry all my belongings with me or should I trust the occasional unknown person across the table to look after my stuff? Nevertheless, with more than 5,500 stores in more than 50 countries around the world, Starbucks shelters on a daily basis thousands of people that make this coffee shop their common workplace.

In fact, nowadays, there are many examples of third places where work dynamics mingle with social activity. Just think about airport lounges, restaurants, malls, etc. But why would people prefer to work uncomfortable in public spaces rather than in an office? What is the reason that makes these people choose these places to work even though security, privacy and connectivity are not ideal? Unarguably, one of the main factors is the growing importance of the feeling of belonging related to what people do for a living. Why they do it, how they do it, and even where they do it defines their inner selves as much as their social selves. Belonging to a community that thinks, behaves and gathers among the same ideals is motivating, inspiring and facilitates the new and unresolved work dynamics.

It is in this context that new typologies such as co-works have emerged as an answer to this need of community-centred way of working. At the same time, firms and offices around the world are increasingly adapting their workplaces to spaces where a shared experience can facilitate work dynamics in an innovative collaborative and inspirational environment.

3. A Three-Dimensional Shared Experience

Community is a broad and flexible concept. Narrowing it down may be short-sighted in many ways, but still offers the opportunity to unveil key concepts that can give important design guidelines. First of all, a community is identified not by its members, but by the elements that link these members together in a common experience. There are three main dimensions in which these linking elements can be set apart.

更具创造性的互动。所有社会都有非正式的会面场所（即'第三空间'），现代社会的新鲜之处在于人们正在努力寻找第三空间。"[1]

我们都同意一点：在星巴克工作并不舒服。那里的插座很少，桌子不稳定而且是圆形（不适合笔记本电脑的形状），并且很容易就会把咖啡洒在电脑或文件上。星巴克也不安静，相反，那里的视觉和听觉干扰很多，没有隐私，去洗手间也不方便：我是该带上随身物品还是请对面的陌生人帮我照看一下呢？然而，在遍布50多个国家的5,500多家星巴克里，每天都有数以万计的人在进行日常办公。

事实上，还有许多其他的可供人们将工作与社交结合起来的第三空间，比如机场休息室、餐厅、商场等。但是与办公室相比，人们为什么喜欢不舒服的公共场所呢？是什么让他们选择了这些缺乏安全感、私密性乃至连通性的第三空间呢？毫无疑问，主要因素之一是人们越来越注重生活的归属感。人们做事的原因、方法乃至场所都决定了他们的内在自我和社会自我。归属于一个集合了拥有相同思想、行为的社区会让人感到欢欣鼓舞，从而帮助人们完成未处理办公活动。

在这种情形下，为了应对人们社区型办公的需求，合作办公等全新的办公类型出现了。同时，全球的公司和办事处都开始调整他们的办公空间，向共享体验靠拢，力求让创新协作、灵感四溢的环境帮助人们更好的完成工作。

3. 三个维度的共享体验

社区是一个广泛而灵活的概念，将其范围缩小可能显得目光短浅，但是这能让我们了解几个重要的设计原则。首先，社区并不等同于社区成员，而是代表着将这些成员结合在一起形成共同体验的各种元素。这些元素可以划分为三个主要的维度。

物理维度： 即社区的物质元素和物理环境，可以是地理邻近场所（街道、居民区、城市）、共用的物品（车辆、家具、玩具等）或景观地标乃至天气。简而言之，"事物让我们聚在一起。"

社交维度： 即通过生活传统和生活方式将我们聚在一起的共同体验，主要表现形式是家庭和朋友，也可以是通过共同爱好聚集起来的一群人。这个维度可以称为"我们共享的故事"。

Physical Dimension: Understood as the material elements and the physical context and surroundings that define a community. It may be geographical proximity (streets, neighbourhoods, cities), use of common objects (cars, furniture, toys, etc.), or even landscape landmarks and weather. In simple words, these are "the things that hold us together".

Social Dimension: Understood as the common experiences that bring us together in tradition and ways of life. Its main manifestation is family and friends, but it can also be activated through behaviour, basically by gathering those who act alike. This dimension can be defined as "the stories that we share".

Identity Dimension: This dimension refers to the world of ideas that define identity. Purpose, epic causes, strong beliefs and profound ideals determine political, religious and social affinity with other people. Having a common cause is a powerful driver for community life. It's "the ideas that define us".

These things, stories and ideas can be identified, measured and interpreted in any community one might observe. Being able to read and understand work communities through these three lectures may allow sensitive interventions that can shelter in a more comprehensive and efficient way the inhabitants of a workplace. Obviously, this is true for any type of community, either that of a business office in New York or that of a market in Korea. Anak & Monoperro are two Spain-based artists who explain their job with the term Urban Animism which they define as constant engagement "in exploring patterns of perception confined in our culture by using time-based mediums and participatory interactions". Visiting the Seoksu Market in Korea, the artists where inspired by a dream in which the Seoksu Spirit appeared personified as wise woman whose mission was to look after the life and destiny of the people of the market. They decided to materialise the spirit for a few days to make contact with the inhabitants of the market and then dematerialise it afterwards. So they made a doll, dressed her up with typical local clothes and presented it to the inhabitants. The people gave offerings to the spirit and it was later buried in the market itself, with a ritual ceremony that gathered the community. What they really did, was creating a myth that was embraced by a community to the extent of becoming part of its cultural tradition. Well-designed interventions can go that far.

4. Community by Design

Culture is the essence that defines any given community, and it is the finest expression of a community's ideals, stories and physical links. Through a reflective creative approach, design is able to shape unique, multidimensional, suggesting and inspiring spaces and objects that give form to a common shared experience.

身份维度：这一维度指的是确定身份的思想世界。目标、历史原因、强烈的信仰、坚定的理想等决定了人们在政治、宗教、社交方面的亲密度。共同目标是社区生活中强大的驱动力。"理念决定了我们是谁。"

这些"事物、故事和理念"可以定义、衡量和解读任意一个社区。通过这三种维度来解读办公社区，设计师可以实现更敏感的设计，为使用者提供更综合、更高效的办公环境。显然，这三种维度适用于任何类型的社区，无论是纽约的商务办公室还是韩国的农贸市场。阿纳克和莫诺佩罗是两位西班牙艺术家，他们用"城市泛灵论"来解释自己的工作，认为自己在不断地"通过实践媒介和参与互动来探索局限于我们文化中的各种感知形式"。在游览韩国的石水市场时，两位艺术家从石水精神中获得灵感，这种精神具体表现为：聪明的主妇以及市场里人们的命运。他们决定将这种精神"实体化"，通过与当地居民的联系和了解，他们做了一个娃娃，给她穿上了典型的当地服装并进行展示。人们给她供奉贡品，然后整个社区聚集起来举行例行仪式，把她埋在了市场里。事实上，这两位艺术家的行为是创造了一个被整个社区所相信的神话并使其变成了当地文化传统的一部分。好的设计所达成的效果就是这么深远。

4. 社区型办公空间设计

文化是社区的基本元素，也是社区理念、故事和物理联系的最佳体现。通过具有反思性和创造性的方式，设计能塑造出独特、多元化、具有启发性的空间和食物，为人们提供共享体验。文化位于物理、社交和身份维度的交叉点，因此，至少可以从三个层面上对其进行处理。

情感记忆（当物理维度遇上身份维度）：事物与空间能有效地吸引某特定社区居民的情感记忆。在规划办公空间时，必须仔细设计和选择物理空间、家具以及配饰，它们必须既实用又有象征意义，能够直接引起特定社区的情感共鸣。简洁和朴实是设计的关键。复杂的未来派设计可能会产生惊艳的效果，但是很难与人们的情感连接起来。阿道夫·卢斯在《山区建筑设计守则》中提到："不要怕人批评你过时……守旧是一种坚持。真相有上千年的历史，它与我们内心的感受仍然保持着最紧密的联系，而谎言如影随形，日新月异，可它终究是谎言。"

交流与启发（当身份维度遇上社交维度）：社区的延续性和

Culture stands in the intersection of all three physical, social and identity dimensions of community and it can be intervened by at least three approaches, defined by the intersection of the mentioned dimensions.

Emotive Memory (when physical meets identity): Objects and space can be powerful components that appeal to the emotive memory of inhabitants of a given community. When configuring a workplace, physical space, furniture and accessories must be carefully designed and chosen to be both practical and symbolic, in a way that appeals directly to those emotional components that define the identity of a community. Simplicity and honesty in design is key. Complex and futuristic design may have an important wow effect, but frequently fails to connect emotionally with people. Adolf Loos better describes it in his "Rules for Those Building in the Mountains": "Have not fear of being chastised as outdated. ...remain with the old. For even if it is hundreds of years old the truth has more connection with our innermost feelings than mendacity, which paces alongside us."

Communication and Inspiration (when identity meets social): Continuity and endurance of a community relies in great measure in its ability to communicate its own story. Passing to others a community's shared experience is inspiring and gives form to new ways in which the community can develop and expand. Rituals, ceremonies, traditions and custom habits of a community have to be observed and reinstalled when designing a physical space for that given community.

Assertiveness is Elegance (when social meets physical): In many ways, a good design is the one that is able to "disappear" under the presence of human dynamics. Work dynamics, as formerly expressed, have been disassociated from physical space, meaning that it can occur anywhere, despite the immediate environment. Assertive, practical interventions that facilitate social interactions and promote new work dynamics are commonly much more efficient than high profile over the top designs that can get lost in pyrotechnical efforts. For Enzo Mari, this condition lies in the heart of any good design:

"The quality-quantity ratio is central to the whole of industrial production: quality is determined when the shape of a product does not 'seem', but simply 'is'. This statement, anything but a paradox, is not however understood by most people. And this makes it particularly difficult to execute projects of real worth."

Cristián Olivi Ianiszewski

持久性取决于它的交流沟通能力。向其他人传递社区的共享体验是具有启发性的，能为社区的发展壮大提供新思路。在为特定社区设计物理空间时，必须要仔细观察并重新安置社区的惯例、典礼、传统和习俗。

自信即是优雅（当社交维度遇上物理维度）：从许多角度来说，一个好设计必须能够在人类进行活动时"消失"。正如上文所提到的，办公活动已经脱离了物理空间，可以发生在任何地点，无需考虑直接环境。自信、实用的设计能促进社交互动，激发新的工作活动，比那些令人眼花缭乱的高调设计更加高效。设计师恩佐·马里认为此条件适用于任何好设计：

"质量－数量比是整个工业生产的核心：质量取决于产品'是什么'，而不是'像什么'。但是大多数人都并不理解这种说法。因此，想要做出真正具有价值的项目就变得愈加困难。"

克里斯蒂安·奥利维·伊安尼塞维斯基

Reference:
参考资料
[1] Wikipedia. http://en.wikipedia.org/wiki/Third_place

GOOGLE ISRAEL OFFICE TEL AVIV
CULTURAL THEMES CREATE MULTI-FUNCTIONAL WORKSPACE

Tel Aviv, Israel

以文化特色为主题打造出多功能办公空间
谷歌特拉维夫办公室 / 以色列、特拉维夫

It is a new milestone for Google in the development of innovative work environments: nearly 50% of all areas have been allocated to create communication landscapes, giving countless opportunities to employees to collaborate and communicate with other Googlers in a diverse environment that will serve all different requirements and needs.

There is clear separation between the employee's traditional desk-based work environment and those communication areas, granting privacy and focus when required for desk-based individual working and spaces for collaboration and sharing ideas.

Designer: Camenzind Evolution, Setter Architects and Studio Yaron Tal　Area: 8,000sqm
Completion Date: 2012　Number of Workstations: 490

本项目是谷歌在创意办公环境开发中的里程碑：近50%的空间被用于打造交流景观，赋予了员工们无限的合作和交流机会，多样化的环境能满足各种各样的需求。

这些交流区域与传统的以办公桌为基础的办公环境有着显著的差别，既能为专注集中的工作提供私人办公桌，又能为合作和意见交换提供交流空间。

设计师：Camenzind Evolution 设计公司、Setter 建筑事务所、Yaron Tal 工作室　面积：8,000 平方米　竣工时间：2012 年　工作台数量：490

Each floor was designed with a different aspect of the local identity in mind, illustrating the diversity of Israel as a country and nation. Each of the themes was selected by a local group of Googlers, who also assisted in the interpretation of those chosen ideas.

Level 27
Theme: Pleasure & Delight
Colour: Bordeaux Red

每层楼的设计都有一个当地形象的主题,展示了以色列国家和民族的多样化特色。各个主题都由谷歌的本地工作人员精心挑选,他们还参与了设计实施的辅助工作。

27层
主题:愉悦与快乐
色彩:波尔多红

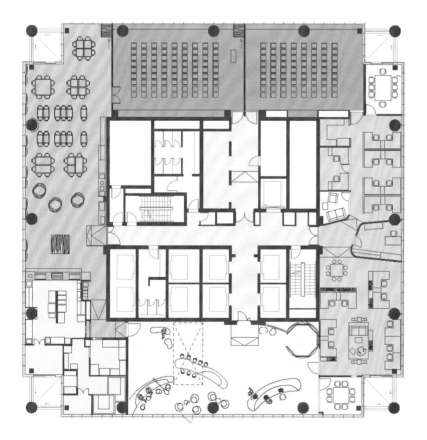

Level 29
Theme: Innovation & Hospitality
Colour: Yellow

29层
主题：创新与好客
色彩：黄色

Community

Level 33
Theme: Culture & Heritage
Colour: Gold & Saffron

33层
主题：文化与传承
色彩：金色与橙黄

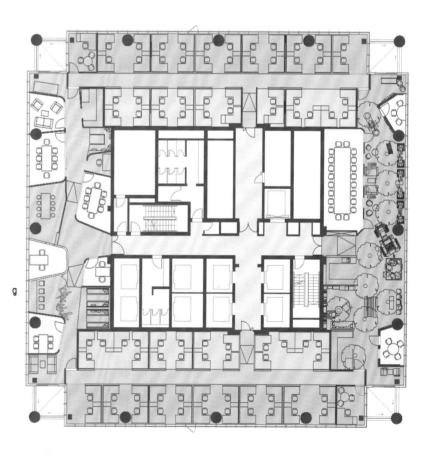

Community

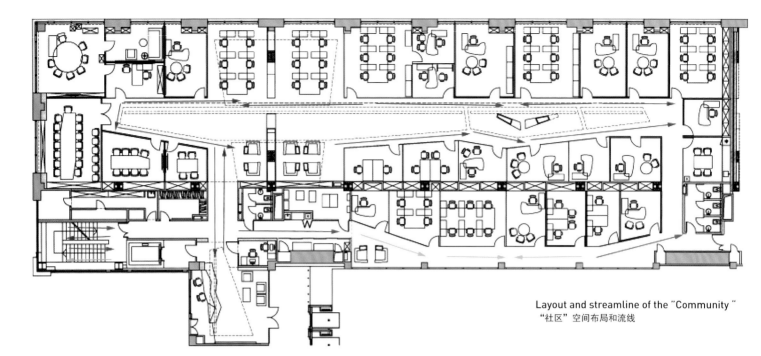

Layout and streamline of the "Community"
"社区"空间布局和流线

COMCAST SILICON VALLEY INNOVATION CENTRE

THE NON-PARTITION RELAXING CENTRE CREATES AN URBAN LIFESTYLE

Sunnyvale, USA

无区隔的休闲区营造城市生活的氛围
康卡斯特硅谷创新中心 / 美国，森尼韦尔

New Innovation office provides both working and playing space for the rapidly growing company that changes how people connect to entertainment, information, and their working.

Needing an innovative and cool space for the Sunnyvale team, Comcast partnered with Design Blitz to create a cutting-edge, collaborative new community for the workers to have fun during the boring working time. The Comcast team is an especially bright and collaborative group, preferring to work and play in an open office environment which they refer to as their "innovation community".

Designer: Blitz   Area: 2,880sqm   Completion Date: 2013
Photography: Jasper Sanidad

全新的创意办公空间为快速发展的公司提供了办公和玩乐的空间，作为美国最大有线系统公司，康卡斯特改变了人们与娱乐、信息和工作的连接方式。

康卡斯特公司的森尼韦尔团队需要一个创新而出色的办公空间，因此他们与Blitz设计公司合伙打造了一个走在时代前沿的协作型社区空间，让员工在工作时间也能尽享乐趣。康卡斯特团队是一直特别欢快而乐于合作的队伍，喜欢在开放式办公环境中工作和玩乐，他们称其为"创新型社区"。

设计师：Blitz 设计公司　面积：2,880 平方米　竣工时间：2013 年
摄影：雅斯佩尔·萨尼达德

In the centre of the "innovation community" locates a new relaxing centre for the staff to play games, giving countless opportunities to employees to collaborate and communicate. There is no clear separation between the employees' traditional desk and those communication areas. The communication areas and the coffee room are located around the relaxing centre just like different small "villages", creating a new sense of urban life where people could walk and talk casually.

"创新型社区"的中央是一个休闲中心,员工可以在此游戏,也提供了无限的交流和合作机会。传统的办公桌与这些交流区之间没有明确的界限。交流区和咖啡室就像不同的小村落一样围绕着休闲中心展开,营造出一种边走边交流的城市生活之感。

Cool, muted colours prevail over the space, with vibrant pockets of red throughout. Reflecting Comcast's origins as a cable provider, the community design of the space takes cues from a two-dimensional electrical wiring diagram. Red paths run across the floor using in the pattern of wires and serve as way-finding guidelines, which also connect different villages. The architects further extrapolated the pattern into a three-dimensional form, creating red hooded structures which divide the space and provide alternate collaboration and meeting areas.

While the new Innovation Centre is tech-centric, it was important for the project to incorporate green technology and be environmentally friendly. Materials with multiple uses were selected throughout, such as ones that had privacy, acoustic, and aesthetic functions. The project is currently seeking LEED Gold certification.

整个空间以清爽、柔和的色彩为主，同时也点缀着一些活泼红色。为了反映康卡斯特是一家作为有线电视供应商起家的公司，空间的社区设计从二维电路图中获得了灵感。红色路径以电线的图案穿过楼面，起到了指路的作用，将各个村落连接起来。建筑师进一步将电路图案拓展到三维造型，形成了红色的建筑结构，既起到了空间分割的作用，又可作为交流、会面区域。

创新中心以技术为核心，因此项目必须融入绿色科技和环保技术。项目所选用的材料大多具有多重功能，如私密、隔音、美观等。项目目前正在寻求 LEED 绿色建筑金奖认证。

Community

Community

HALLE A

CREATING A HOME-LIKE ATMOSPHERE WITH "VILLAGE" AS ITS CORE ELEMENT

Munich, Germany

以"村庄"为核心元素营造居家办公的氛围
A厅办公空间 / 德国,慕尼黑

Community working environment is a space which immediately creates the desire in anyone entering it – client, colleague or supplier – to explore the place and become a part of it. It also creates a new working environment for the staff in which work once again becomes solidly grounded and creativity can be a natural part of life. The design concept was inspired by the dynamic interaction between architectural atmosphere, the modern digitally focused world of work and the contemporary search for intimacy and material substance.

Elements saved from the hall's days as a machine shop, such as the large industrial clock, add deliberate references to the industrial age and its core values of rationality, efficiency and entrepreneurial spirit that play such a vital role in creating the atmosphere.

Echoes from the past age of industrial labour blend with the spirit of work in the modern digital age. The natural roles of players within a global network, the permanent state of being "connected" and the shift from physical production to abstract media have now become a matter of course for all staff. The direct physical confrontation between industry and digitalisation hones an awareness of problems such as a loss of direct, "authentic" experience and a growing longing for substance and personal commitment. Restoration of closeness may be an answer. The "village" is the core element in the design of the hall, generating identity and serving as a symbol of manageable scale, proximity and personal involvement.

The ridge-roofed houses, the village square with park bench and the open-plan working areas in the "garden" around the brass-walled house create a grounded, tranquil atmosphere.

Designer: Designliga   Completion Date: 2013

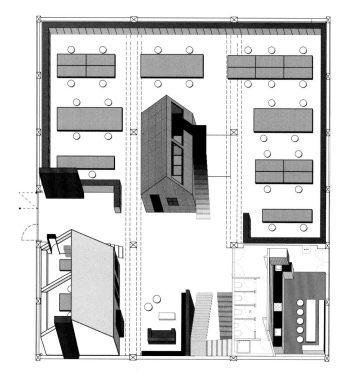

Layout of the "Community"
"社区"空间布局

社区办公空间能迅速使人产生探索并称为空间一部分的欲望,无论是客户、员工还是供货商。它还能为员工打造全新的办公环境,让工作变得脚踏实地,让创意成为生活的一部分。

项目设计概念的灵感来自于建筑氛围、现代的数字化办公以及现代人对亲密感和材料质感的追求。

办公空间的前身是一家机械工厂,因此设计师保留了大型工业时钟等怀旧元素,给人以工业时代的感觉,突出了工业时代的理性、效率、进取精神等核心价值,以此来营造办公氛围。

呼应工业时代的元素与现代数字时代的办公精神被融合在一起。作为全球网络的一员,公司与世界的"连接"以及从实体生产向抽象媒体的转化是全体员工的重要课题。工业与数字化之间的直接对抗也产生了一些问题,例如,直接、真实体验的缺失,对实物和个人承诺日益增长的渴望等。重塑亲密感就是解决问题的关键。大厅的设计以"村庄"为核心元素,它生成了身份感,充当起可管理尺度、亲近感和个人参与的象征。

高脊房屋、配有公园长椅的村庄广场以及房屋周围"花园"里的开放式办公区营造出一种真实、宁静的氛围。

设计师:Designliga 设计公司　竣工时间:2013 年

设计的出发点
大厅呈长方形,天花板的最高处可超过 10 米。东西两侧采用玻璃砖建造,在齐眼的高度有一排窗户,让整个大厅都洒满了柔和的光线。楼面由斜纹落叶松木地板和混凝土两种材料拼接而成,增加了视觉结构。裸露的起重轨道和取暖装置遍布整个大厅。室内的部分空间被分割成两层;前工长的办公室俯瞰着大厅和入口,入口两侧是红色的砖墙。

原有特征和变化
机械工厂的楼面、砖砌结构和起重轨道得到了保留,仅进行了清洁和涂漆处理。两座双层高的高脊"房屋"矗立在空间的正中,形成强烈的对比。作为空间规划概念的主要元素,它们分割出独立的区域,赋予了四周的大厅空间结构感。"房屋"具有多重功能,员工们可以自由放松、随便闲聊或是进行保密工作。设计师把"社区概念"带进了这些"房屋"里,形成了轻松的办公氛围。

Starting-point
The hall has a rectangular footprint and ceilings over 10 metres in height at their highest point. The east and west sides are built of glass bricks and incorporate a row of windows at eye level. The hall is thus flooded with a consistent level of glare-free light. Flooring combines areas of cross-grained larchwood parquet with concrete, adding visual structure to the hall. Exposed crane tracks and heating elements extend throughout the entire hall. Part of the interior is two-storey; what was once the foreman's office overlooks the hall and the entrance, which is flanked by two brickwork walls.

Existing Features and Changes
The floors, brickwork and crane tracks from the former machine shop were retained and merely cleaned and painted. A pair of two-storey ridge-roofed "house" enclosures now stand at the heart of this atmospheric setting, forming the most striking architectural contrast. A key component of the space planning concept, they establish distinctive areas and give structure to the hall space around them without the need for formal boundaries. The "houses" are used by staff involved in extensive relaxing, casual talking or working with confidential information. The designer takes "Community Concept" into these "houses" to create a more relaxing working environment.

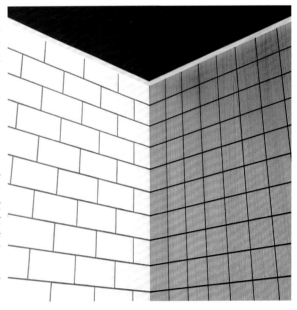

The two foreman's offices on the upper floor were converted into conference rooms with contrasting styles – a functional room with classic conference equipment, and a salon for informal meetings. The solid wall between the two offices was removed and replaced with glass. The kitchen and toilets are situated on the ground floor immediately underneath the conference rooms. A steel door separates these private areas from the hall's work area.

The basement houses a model-building workshop, a screen-printing press and a photographic studio and can be viewed through a glass section in the floor.

二楼的两间工长办公室被改造成两个风格截然不同的会议室——一间配有典型的会议设施，另一间则为用于非正式会面的沙龙。两间会议室间的实体墙被拆除，由玻璃取而代之。厨房和洗手间都设在一楼，就在会议室的下方，一扇铁门将这些私密空间与大厅的办公区隔开。

地下室里设有模型制作室、丝网印刷机和摄影工作室，从一楼地面上的玻璃板就能看到下方的景象。

The area between the two houses forms a "village square". This point is the intersection of both axes of the interior: from entrance to conference rooms, and from workspaces to relaxing space. The "square" extends into an open-plan library area.

两座"房屋"之间的区域是"村庄广场",这里是室内空间两条轴线——从入口到会议室,从办公区到休闲区——的交汇点。"广场"一直延伸到开放的图书阅览区。

Workspaces are distributed around the outside walls of the central brass-finished "house" like the scattering villages. The "house" is the relaxing centre of this community. They are bordered by an 85-metre length of sideboard, which serves as a reception desk at one end and separates work areas from the circulation space at the other.

办公空间围绕着"铜屋"的外墙展开,就像散落的村庄。"房屋"相当于整个社区的休闲中心。它们被85米长的侧板包围,这块侧板的一端形成了接待前台,另一端将工作区与流通空间隔开。

The kitchen is equally important in establishing a feeling of connection and is used for personal free time, together and individually. It features homely elements including a large dining table, individually grouped chairs, lamps and refrigerator with rounded, appealing lines, and a window overlooking green spaces, all serving to unite the "villagers" in a familiar and intimate community.

村庄在建立连接感的过程中同样重要,供私人独自或集体使用。它拥有大餐桌、独立的系列座椅、灯具、可爱的圆角冰箱以及俯瞰绿色空间的大窗,充满了居家氛围,所有设置的目标都是让"村民们"在亲切熟悉的社区中联合起来。

Community

PART 4

MOBILITY – SETTING OF NON-REGIONAL WORKING BELT

第 4 章 流动型办公空间——非区域性工作带的设置

MOBILITY IN ARCHITECTURE

Nowadays, architectural design procedure is becoming more and more demanding, not only in terms of technical skill, but also in terms of architect-client interaction. In the past, architects used to design taking into account only their own perception of space and people's needs. As years pass by, people have the increasing will to become part of the design process, to take decisions themselves, to live and work in custom-designed spaces that fit their particular needs.

This is a global trend, and the architectural world should take it seriously into account in order to create user-friendly architecture. The only way to succeed in achieving the architect-client interaction mentioned above is through continuous mobility at all levels, starting from the very beginning, that is, the first meeting with the client.

All mobility steps need to be recorded by the design company, allowing the company to evolve their way of working, from creating simple architecture to creating collective concept. Collective concept is an interactive experience organised in the following steps, in which the entire design team as well as the client is involved from the very first moment of cooperation. For the creative design team, collective concept is a kind of powerful spirit that enables synthesising each client's personality profile, lifestyle and needs, in order to find the perfect design solution for each case. It is a bottom-up procedure, enhancing mobility among engineers of different disciplines (architect, civil, mechanical, etc.) and clients. Future users are closely following the design procedure, having the opportunity to share their ideas and influence the final result.

If the aforementioned procedure seems a theoretical approach, it is time to mention its practical implementation, that is, the way mobility process is reflected on design: mobility turning into space in a handy manner. If mobility and cooperation are main principles of architecture, why not make this evident from the client's first impression of the company? The design of the architectural office or headquarters is of great significance to make people open-minded, ready to co-operate, willing to exchange ideas, knowing that their opinion and way of thinking matter.

"Mobility" is the core concept of communication and cooperation in office

建筑中的流动性

现在，人们对建筑设计方法的要求越来越高，这不仅针对专业技术，还包括建筑师与客户的互动。在过去，建筑师在设计中只需要考虑他们自己对空间的理解和人们的需求。现在，人们越来越多希望参与到设计流程之中，自己做决定，然后再根据自己生活和工作的特殊需求打造特别的办公空间。

这是一个全球趋势，建筑业必须严肃考虑这一问题，以打造人性化的建筑空间。实现上文所提到的建筑师与员工互动的唯一途径是建立多层次的连续流动性，从设计之初第一次与客户见面开始。

设计公司必须记录下所有的流动步骤，形成自己的工作方式，无论是进行简单的建筑设计还是复杂的集合概念。集合概念是由以下的步骤组成的互动体验，整个设计团队以及客户都将在一开始就参与到合作之中。对创意设计团队来说，集合概念是一种强大的精神力量，它能综合每个客户的个性介绍、生活方式以及需求，从而为每个项目找到最完美的设计方案。这是一个自下而上的流程，能够提升不同领域工作人员（建筑师、市政规划师、机械工程师等）和客户之间的流动性。未来用户将紧跟设计流程，有机会共享自己的观点并影响最终的结果。

如果上文提到的流程看起来像一种理论方法，那么下面将讲到它的实践过程，即流动型进程在设计中的反映：流动性能让空间变得便捷。既然流动和协作是建筑的主要原则，何不将其展示在客户对公司的第一印象中呢？建筑事务所办公室或总部的设计对解放思想、促进合作、鼓励交流、达成共识有着至关重要的作用。

流动性是办公空间设计中交流与合作的核心概念。首先，客户需要通过一扇友善的大门进入公司。这扇大门能通往一个面向各种活动开放的空间：展示、团队合作、会面、研讨会以及其他一切可以想象的活动。一个能打造各种层次隐私感的灵活空间是流动效率的关键元素。具有多层次功能的机构必须具有简洁明晰的功能规划，同时还要兼具一种象征意义：

design. Starting from this point, the client may enter the company's world through a welcoming gate design. The gate could lead to such a space designed to be open to all kinds of activities: presentations, group working, meetings, workshops and anything else one could imagine. Flexibility of space to create different levels of privacy during work is a key element to make mobility efficient. Organisation of functions in different levels through a gradual course follows a neat functional diagram while creating a symbolism: space gets more private as you go upwards. It is also essential to foresee different types of mobility according to the level and state of the design process, which is indicated by furniture layout: working in separate desks, sitting around a table or opposite to each other imply a different way of communicating.

In the other way, the whole working space may also represent a mobility network with clients. Display of previous project models and photos in the office can serve as a connection between future and past clients, showing how other people's needs were taken into account and creating a climate of confidence and intimacy. Space expresses a design stable and sincere, like the relationship of the company with clients.

Finally, mobility in architecture is a huge challenge, which should be pursued, not avoided, even if it seems hard to be successfully established in the first place. Mobility can boost the evolution of design and have a great impact on how we perceive human-oriented architecture.

Margarita Skiada
Fellow Architect of Golden Ratio Collective Architecture

越向上走，空间的隐私程度越高。根据设计流程的层次和状态来预见不同类型的流动性也是必不可少的，前者可以体现在家具布置上：在独立的办公桌前工作、围坐在桌子四周、相对而坐都暗示着截然不同的交流方式。

此外，整个办公空间还需要向客户呈现一个流动型网络。在办公室里展示之前的项目模型及照片有助于与客户建立起紧密的联系，能够展示公司是如何考虑到其他客户的需求，在公司与客户之间建立起信任、亲密的氛围。空间能展现设计的稳定和诚意，就像公司与客户之间的关系一样。

最后，建筑中的流动型设计是一个巨大的挑战，虽然很难一下子就取得成功，我们仍然应当尽力实现，而非绕开这个问题。流动性能促进设计的进化，对人性化建筑的发展有着重大的影响。

玛格丽塔·斯齐达
Golden Ratio联合建筑事务所资深建筑师

1305 STUDIO

SCATTERED BOXES ENABLE FREE WORKSTYLE

Shanghai, China

散落的盒子令自由的工作形式成为可能

1305 工作室 / 中国，上海

Multi-Creative Function Office Space with Boxes

The space acts not only for architecture, interior or graphic design, but also for lots of other purposes, fashion show, art exhibition, cocktail lounges, professional lectures, etc. The concept of this design was inspired by a glass cup. The temporary living space is just like a glass cup without the so-called "space border", but with many limitless possibilities. Just as "a glass of water", when it falls into the water, on the contrary it will get limitless possibilities. Putting different substances into the cup will have a new name, may it be "a glass of milk, juice, beer" or others.

The language of the space is the box. Boxes lying around, either piled up or separated freely, and it makes the space of unlimited possibilities. People in the space can easily find a way to adapt and communicate properly, and share joy with others. The wooden box, which is made of oak, has storage capabilities. Through repeated calculations, designers can design the boxes of accurate size. The accurate size enables the space to show different functions. With the interest of Shanghai Lane Culture, the architects tried to make this 300sqm space into a harmonious combination of tradition and modern.

Designer: 1305 STUDIO   Corporate Identity Design: Shen Qiang   Area: 306sqm
Completion Date: 2014   Photography: shen-photo.com

由盒子构成的多功能创意办公空间

项目空间不仅是建筑、室内或平面设计，还拥有很多其他的功能：时装秀、艺术展览、鸡尾酒会、专业讲座等。设计的灵感来自于玻璃杯。临时的生活空间就像是没有所谓"空间边界"的玻璃杯，充满了无限可能。就像"一杯水"，当它掉进水里，它却获得了无限可能。不同的物质放进杯中都获得一个新名字，如"一杯牛奶""一杯果汁""一杯啤酒"等。

空间的语言是"盒子"。盒子散落四处，有的堆叠起来，有的独立存在，它赋予了空间无限可能。空间里的人们能轻易找到适应和交流的方式，与他人共享欢乐。橡木盒子拥有储藏功能。通过反复计算，设计师最终设计出的盒子具有精准的尺寸。精准的尺寸让空间得以展示不同的功能。带着对上海弄堂文化的浓厚兴趣，设计师力求将这个300平方米的空间改造成传统与现代的完美融合。

设计师：1305 工作室   企业形象设计：申强   面积：306 平方米   竣工时间：2014 年   摄影：shen-photo.com

Mobility

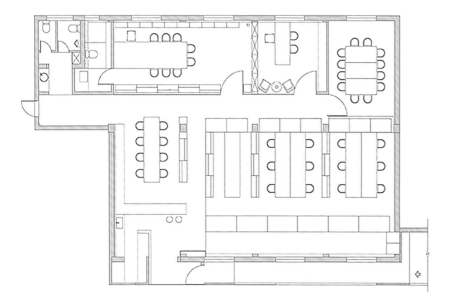

Workspace
Workspace is just like a cup of water, when we work in this "cup", we can see the so-called desks, file cabinets, etc. But for both the desk and file cabinets, designers designed space accurately, so that they could freely combine, to meet the demands over the next five or ten years.

办公空间
办公空间就像一杯水,当我们在这个"杯子"中工作时,我们会看见办公桌、文件柜等。但是在办公桌和文件柜的设计中,设计师对空间的设计十分精准,使得它们可以自由组合,能够满足公司在未来 5 年乃至 10 年时间内的需求。

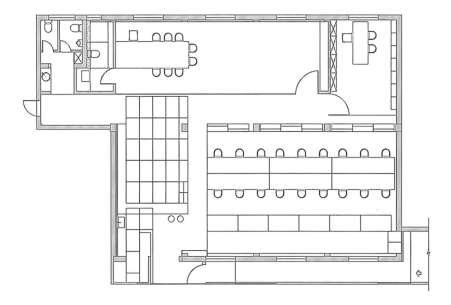

Party Space
Every weekend, the office space becomes an entirely different place. The front desk would be used as a cool DJ counter, or a counter with all kinds of drinks, with a 7-metre-long screen showing various psychedelic images.

派对空间
每个周末,办公空间将进行彻底的变身。前台会变成 DJ 台或饮品台,7 米长的超大屏幕将展示各种迷幻的影像。

T Station
The T station is designed with wooden boxes. At weekends, people can hold a small fashion show together with friends.

T台
设计师用木盒子来打造T台。周末,人们可以与好友们共同举办小型时装秀。

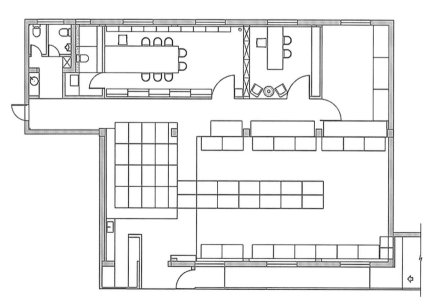

Art Exhibition

At weekends, when the office space is no longer a working space, it turns into an exhibition hall or an art exhibition.

艺术展览

周末,办公空间不再是工作场所,而将变身为展览厅或举办艺术展览。

Mobility

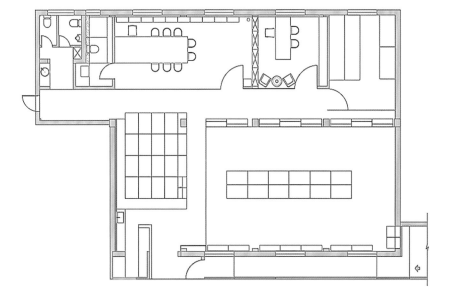

Library

When all the bookshelves are folded, it turns into small library. The heights of wooden boxes are 15cm and 30cm. They can be freely combined. They could be superimposed to 45cm, 60cm, 75cm or 90cm and so on. The shelves could be divided into units with a width of 15cm and 30cm, to increase stability, while in the meanwhile, in consideration of circulation, environment protection, and sustainable development, the traditional lockers and bookshelves could be divided into small units.

图书馆

把所有书架叠起，就成为了小型图书馆。木盒子的高度为15厘米和30厘米。它们可以自由组合和叠加，形成45厘米、60厘米、75厘米、90厘米等高度。书架可以被进一步被划分成15厘米和30厘米宽的单元，相互交错，以增加稳定性。同时，考虑到流通、环保、可持续发展等因素，传统的储物柜和书架将被划分成更小的单元。

Mobility

Lecture
Two white wooden supports of different modulus can turn into a speech sitting stool. It could meet a small speech with about 50-100 people.

讲座
两个白色的木质模块支架可以变成一个听讲座用的小凳。整个空间能举办 50~100 人的小型讲座。

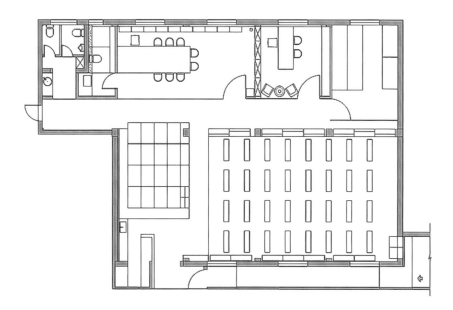

Mobility

THE BRIDGE

USING A "BRIDGE" TO CONNECT DIFFERENT AREAS

London, UK

以"桥梁"作为各区域的连接元素

桥梁办公 / 英国，伦敦

A Bridge in Work Environment Makes Better Circulating

Bridging between floors, this elegant connecting element encourages interaction and circulation between employees and creates pockets of space in which to work and gather. Designed to be a flexible environment which unites four companies at a new HQ, it forms an inspirational and circulating work environment. The bridge can create an impressive entrance, strategically connect employees, departments and companies, form a variety of new spaces to meet and collect within and create a beautiful new café and event space for both employees and visitors to enjoy.

A 64m long, undulating, multi-level structure spanning two floors in a double-height void within 1,500m² of office and recreational spaces. Conceived as a continuous folded surface, the Bridge is constructed from pre-fabricated cross laminated timber (CLT), and is a structurally dynamic form, spanning over 8m at a time.

Designer: Threefold Architects   Client: Bathroom Brands   Completion Date: 2014

用桥梁在办公环境中打造更好的流动性

作为一个优雅的连接元素，楼面之间的桥梁鼓励员工之间进行互动和交流，同时也形成了若干个可以办公和集会的口袋空间。为了将四家公司聚集在一个新的总部，这个灵活的办公环境必须具有启发性和流动性。桥梁所打造的入口令人印象深刻，有效地连接了员工、部门和公司，形成了各种新鲜的空间来进行会面，并且还为员工和访客打造了漂亮的咖啡厅和活动空间。

在 1,500 平方米的办公空间和休闲空间里，一个 64 米长、上下起伏的多层结构横跨两个楼层。作为一个连续的折叠结构，桥梁由预制交叉层压板建成，其结构富有动感，跨度可达 8 米多。

设计师：Threefold 建筑事务所   委托方：卫浴品牌公司   竣工时间：2014 年

Mobility

An Impressive Entrance
Both staff and guests enter into a light, airy space where the CLT Bridge structure begins. Here the Bridge creates a dramatic first impression – forming a seating area, a staircase and an impressive 5m high wall, then wrapping upwards to the first floor level.

令人印象深刻的入口
员工与访客一走进这个轻盈的空间,桥梁结构就会映入眼帘,它给人们带来了难忘的第一印象,依次形成了座位区、楼梯间和5米的高墙,然后缓缓地蜿蜒上升到二楼。

Mobility

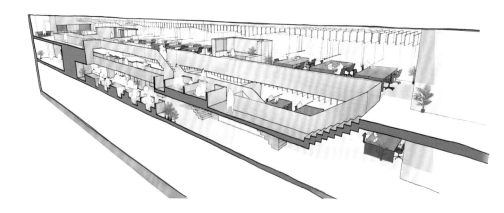

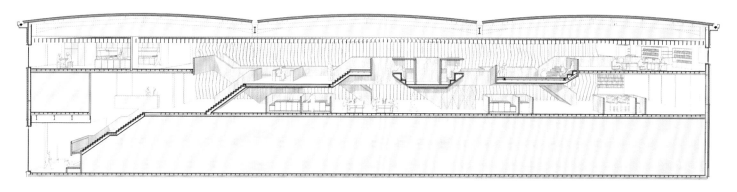

An Elegant Connection

The Bridge continues to the first and second floors. A connecting element between spaces, it encourages horizontal and vertical movement across the office. This connection is key in addressing the notion of community within the building – bringing together the different departments and companies at strategic points.

优雅的连接

桥梁继续延伸到二楼和三楼。作为一个连接元素,它能促进办公区域内在水平和垂直方向上的运动。这种连接性是建筑社区概念的关键所在,能将不同的部门和公司聚集在特定的节点上。

A Collection of Meeting Places
Inspired by historic inhabited bridges, the folded CLT structure is sculpted to form spaces above, below and within, for interaction and gathering. These areas for interaction vary in size from 1-2 person booths to a 40 person forum.

会面场所的集合
设计从传统的栖居式桥梁中获得了灵感，折叠的桥梁结构在上方、下方和内部都分别形成了空间，用于互动和聚集。这些互动空间的尺寸从可容纳1~2人的格子间到可容纳40人的论坛空间，大小不一。

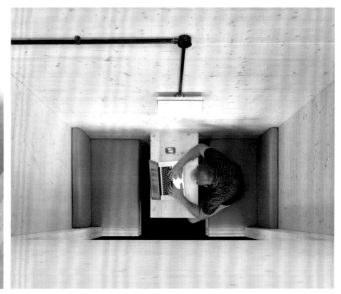

A Bustling Café

Conceived as an extension of the Bridge, here the folded timber surface forms counters, storage and booth areas. The café is a light-filled space with full height glazing and a large balcony. It is a place for meeting and eating, a place to take clients, and to hold company events.

忙碌的咖啡厅
作为桥梁的延伸，折叠的木板平面形成了台面、储藏间和卡座区。咖啡厅配有落地玻璃窗和大阳台，宽敞明亮，是休闲会面、就餐、与客户商谈、举办公司活动的好去处。

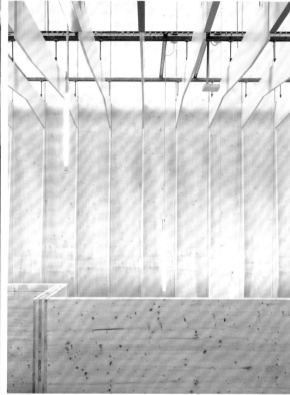

A Dynamic Backdrop

The rear wall and ceiling above the bridge form the backdrop to the office space. On this blank canvas the designers created a 48m long installation of gently undulating fins to the wall and ceiling. Roof lights over this space bring in a soft natural light, which is enhanced by a series of long ethereal pendant lights. The delicate waves of the fins and lights bring to mind the water that the bridge passes across, creating a dreamlike space within the office environment.

富有动感的背景

后墙和桥梁上方的天花板形成了办公空间的背景。在这块空白的画布上，设计师打造了一个48米长的波浪形翅片装饰装置。头顶的天窗带来了柔和的自然光，而一系列垂吊下来的轻盈吊灯更是增强了这种感觉。翅片和吊灯的精致波浪令人想起了桥梁，营造出梦幻般的办公环境。

ORANGE GROVE ATHENS

THE MOBILE FURNITURE ENABLES CHANGEABLE SPACE

Athens, Greece

设施的可移动性打造空间的可变性

雅典橘林产业园 / 希腊、雅典

Three Plans and Mobile Furniture

Orange Grove Athens is a start-up incubator for young entrepreneur, organised by the Dutch Embassy in Athens. The concept of mobility, flexibility and connectivity was defined by them as they are the qualities they want the young entrepreneurs to obtain through the activities in the incubator. Architects translated this concept into an architectural design that is why architects made a fully flexible layout for the office space where all furniture and even some closed spaces such as the meeting space are mobile. This means the space can adapt to many types of events and architects hope it inspires the users to be creative about the space and therefore lead to creative cooperations with other entrepreneurs in Orange Grove.

The three plans show this clearly: in the regular layout all furniture is in its place; here the orange circles, which refer to the oranges (this is a Dutch concept, because Orange is the national colour) organise the space. The flexible layout shows how architects hope that at the end of the day new cooperations have emerged through the use of the space by the young entrepreneurs. The presentation layout is an example of how the space can adapt to the functions, such as presentations, seminars, lectures up to parties.

Designer: gfra architecture - Tasos Gousis, Joost Frijda, Eddie Roberts, Nota Tsalta, Fotini Anagnostou Graphic Designer: Dafni Evangelou Mechanical Engineer: Technomech - Ch. Milionis, Z. Tsiatsikas Client: Netherlands Embassy in Athens Area: 450sqm Completion Date: 2013 Photography: Costas Lakafossis

三种方案与移动家具

瑞典橘林产业园专门为年轻创业者提供服务，由荷兰驻雅典大使馆组织。他们提出的"移动、灵活、连通"的概念正是他们希望年轻创业者所具备的品质。建筑师将这一概念转化到建筑设计中，打造了一个完全灵活的办公布局，所有家具，乃至会议区等封闭空间都可以自由移动。这意味着空间可以适应各种类型的活动，建筑师希望用户可以以创新的方式使用空间，然后与橘林产业园中的其他企业实现创造性的合作。

三种方案实际非常简单：在常规布局中，所有家具都各归其位，位于橙色的圆圈内（橙色是荷兰的国家代表色）。在灵活布局中，建筑师希望看到一天结束后，年轻企业家通过空间的利用实现新的合作。展示布局则显示了空间如何适应各种功能，例如展示会、研讨会、讲座、派对等。

设计师：gfra 建筑事务所——塔索斯·古希斯、约斯特·弗里贾达、埃迪·罗伯茨、诺塔·萨尔塔、弗蒂尼·阿纳戈诺斯拓 平面设计：达夫尼·埃万耶卢 机械工程师：Technomech 公司——CH·米利奥尼斯、Z·斯亚斯卡斯 委托方：荷兰驻雅典大使馆 面积：450平方米 竣工时间：2013年 摄影：科斯塔斯·拉卡福西斯

Mobility

Move Me
"移动我！"

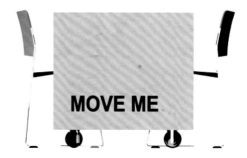

Mobility

Mobility is expressed by making all elements mobile. All furniture is positioned on wheels and even the closed meeting space and concentration spaces can be moved around using the especially positioned track.

移动性

移动性体现在所有的元素都是可移动的。所有家具都配有转轮,甚至连封闭的会议空间和集中空间都能通过特殊的轨道进行移动。

Connectivity

Connectivity is the direct consequence of the layout. In order to work on a specific spot or in a specific layout one just pushes the furniture in place, plugs in the power cable to the wall socket and connect to the world.

连通性

连通性是空间布局的直接结果。如果想在特定的场所或布局中工作，人们只需把家具推到固定的位置，插上墙上的电源，就能与全世界相连。

Connect Me
"连接我！"

Join Me
"加入我！"

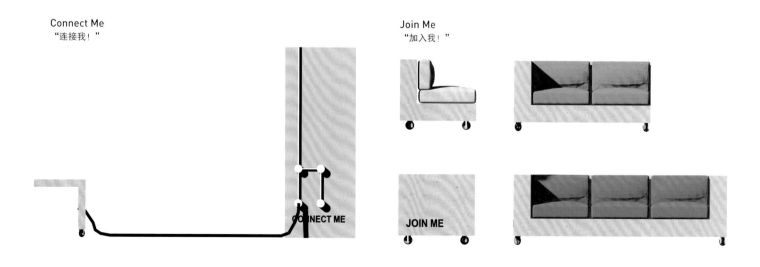

Organize Me
"组织我！"

Meet Me
"遇到我！"

Hide Me
"隐藏我！"

Share Me
"分享我！"

 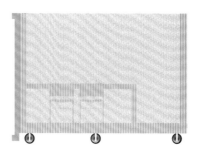 

Mobility

Flexibility

Through this mobility full flexibility is achieved and the space is fully adaptable to any function. During the day the working space can be transformed to allow any form of cooperation. And with the help of flexible space dividers, at any point of the space, a separate area can be created, for presentations or meetings, up to clearing all furniture out of the way for a conference of 250 people.

灵活性

移动性实现了空间的灵活性,使其可以适应各种功能配置。白天,办公空间可以任意转化,适应各种形式的合作。在灵活空间隔断的帮助下,可以在任何地点随意打造出独立的空间,用于展示、会议等。在清空所有家具后,整个空间可以举办容纳250人的会议。

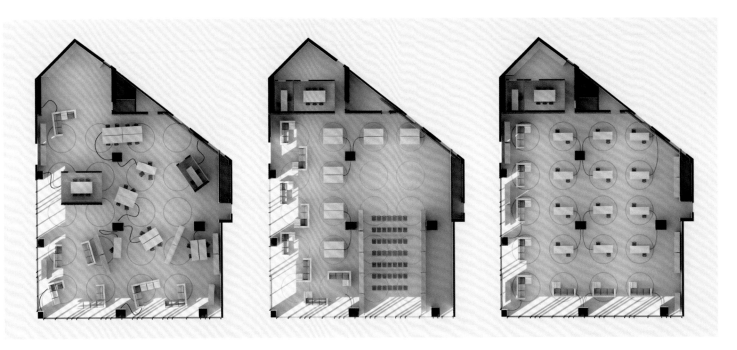

Changeable Space
可变的空间

Over Plan
总平面

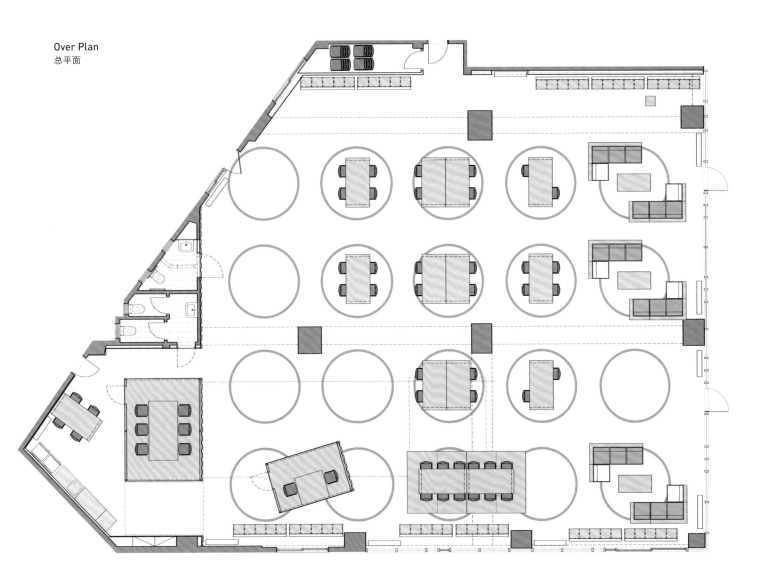

Mobility

JUSTPEOPLE
INTEGRATION WITH URBAN PUBLIC SPACES

Santiago, Chile

城市公共空间的巧妙植入

唯人公司 / 智利，圣地亚哥

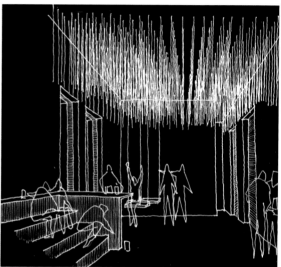

JustPeople proposes an opportunity to meditate about the value of the encounter of people and ideas in a collaborative work experience, both positive and inspiring. This experience is designed for an intelligent, curious, reflexive and enthusiastic adult. The place offers a constant discovery of objects, artifacts, visual hints, small surprises and smooth details. The sober, black-painted environment enhances the complexity and the importance of the interactions between people, with the space, the objects, the scents, the flavours and the ideas. The parts of the layout, diverse on its nature, and specific on its functional requirements, are distributed in a two-storey regular transparent volume which provides direct street access and natural sunlight for almost all the indoor spaces.

Architect: Cristián Olivi Design Team: Daniela Reyes, Diana Menino, Roberto Gutierrez Identity Design: Simplelab Construction: Contract Workplaces Furniture Design: Manuel Oneto Special Furniture: Los Gogo Lighting Design: Docevolts Special Lighting: SimpleLab + We say Materials & Equipment: Microfloor, multicarpet, rollux, Ducasse, Spatti Infographics: Daniela Reyes, Cristián Olivi Sketches: Roberto Gutierrez Lamp Process Photographs: SimpleLab Area: 1,100sqm Completion Date: 2014 Photography: Nicolás Saieh

1. Access control	7. Restroom	1. 门禁	7. 卫生间
2. Public access	8. Kitchen	2. 公共入口	8. 厨房
3. Coffee house	9. Services	3. 咖啡屋	9. 维修室
4. Coffee bar	10. Foyer	4. 咖啡吧	10. 门厅
5. Cashier	11. Foyer bar	5. 款台	11. 大堂吧台
6. Host	12. P lay lab	6. 主控制室	12. 游戏室

唯人公司的办公空间设计让人重新思考了人与人相遇的价值以及合作办公体验的概念，积极而富有启发性。这种体验专为聪明、好奇、积极、热情的成年人所设计。人们在空间内会不断发现新的目标、物品、视觉暗示、小惊喜和流畅的细节。严肃的黑色环境提高了空间的复杂性，突出了人与人以及人与空间、物品、气味、味道、想法之间的互动。空间布局的各部分各具特色，分别满足不同的功能需求。这个两层高的透明空间与街道直接相连，几乎每个角落都能享有自然采光。

建筑师：克里斯蒂安·奥利维　设计团队：丹妮拉·雷耶斯、戴安娜·梅尼诺、罗伯托·古铁雷斯、SimpleLab 工作室　施工：Contract Workplaces 公司　家具设计：曼纽尔·欧内托　特殊家具设计：洛斯·戈高　照明设计：Docevolts 公司　特殊照明设计：SimpleLab 工作室 + We say 公司　材料与设备：Microfloor 公司、multicarpet 公司、rollux 公司、Ducasse 公司、Spatti 公司　信息图形设计：丹妮拉·雷耶斯、克里斯蒂安·奥利维　草图：罗伯托·古铁雷斯　灯光进程摄影：SimpleLab 工作室　面积：1,100 平方米　竣工时间：2014 年　摄影：尼古拉斯·塞耶

Ground Floor Plan
1 层平面图

Mobility

Coffee-based Dynamics
The public and (coincidentally) most crowded-like dynamics such as those of the Coffee House and the Labs are distributed in the ground floor. The Coffee House is street oriented, seeking the visibility needed to serve as a referent, an invitation. It extends its influence to the exterior with a comfortable terrace which is both an extension of the Coffee House and an urban public space. A long, wide Foyer articulates the relation between the Coffee House and the Labs. It serves as a mediator between the more casual, sensory Coffee-based dynamics, with those that occur on the Labs, which tend to be more exploratory and reflexive. These idea-based Play Labs are equipped with functional interventions, such as continuous whiteboard/shelves and mobile space-dividing panels that allow multiple configurations for the two main Spaces.

源自咖啡的活力
咖啡屋、实验室等人来人往的公共空间被设在一楼。咖啡屋朝向街道，拥有不错的视野，同时也能吸引过往的行人。舒适的露台让咖啡屋的影响力延伸到了室外，形成了一个城市公共空间。又长又宽的门厅将咖啡屋和实验室连接起来，在休闲、感性的咖啡屋和具有探索性的实验室之间形成了过渡。以思想为基础的游戏实验室配有各种功能设施，例如连续的白板/书架、可移动式空间隔断（可实现两个主空间的多重配置）等。

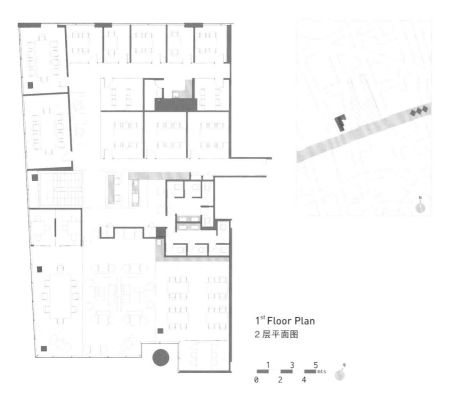

1st Floor Plan
2层平面图

Lounge-based Dynamics

Same as the ground floor, the first floor is configured among the interaction between two diverse but complementary parts of the project. Heading south, overviewing the terrace and the street, the Lounge is an open plan collective office, where furniture and people give form to diverse single, collective, private and casual work dynamics. To the north, on the other hand, 12 project/meeting rooms are distributed, allowing group work for 2, 4, 6 or even 10 people in a comfortable, illuminated, equipped work space.

源自休闲空间的活力

与一楼相同,二楼的空间配置也围绕着两个不同却互补的部分展开。休息室朝南,俯瞰着露台和街道,采用开放式办公布局,利用家具和人来营造一种多样化的办公环境。另一方面,北侧设置着12间项目室/会议室,为2、4、6乃至10人的团队提供了舒适、明亮、设备齐全的办公空间。

Mobility

OoO PRESS2
FREE CHOICES BETWEEN COMPARTMENTS AND COMFORTABLE WORKSPACE

Eindhoven, The Netherlands

格子间与舒适办公的自由选择
PRESS2 办公空间 / 荷兰、埃因霍温

Out of Office appeared to be the embodiment of better cooperation, freedom, suiting the work ethic of the 21st century. It offered a variety work settings such as a garden including a cuddly white rabbit, a comfy work-bed in which one could doze off in between e-mails, a coffee bar, a living room and a library area. The staff could choose any space to do their work and this create a kind of flow-working space.

In the design, studio KNOL responded to the new working trends by creating a variety of flexible and high-efficient environments in the open space of 300m², allowing workers to choose the setting they felt most comfortable. They could change their working space at all times. Out of Office hosted five characteristic areas that – by now – people all recognise as a workspace: a bed, a living room, a garden, a library and a coffee bar. The designer Anna: "We didn't want to create a completely non-work environment (such as home); we don't want an environment that screams work either. We created a space that flows work."

The workers that came for the cubicle office, were actually very satisfied and productive. "We can conclude that flow-working is something very personal; most people value the freedom in their work a lot. Also, they are easily satisfied with the basics for a flexible day plus some creative comfort," says Joep. Celine de Waal Malefijt. "Out of Office was an art project and by no means a full-on scientific experiment. It is a story in which we took people along. By experimenting with the effects of social and architectural design on work performance and satisfaction, we hope to have contributed to a better understanding of it."

Designer: studio KNOL (Jorien Kemerink & Celine de Waal Malefijt) Design Collaboration: Christiaan Bakker Social Research: Anna Dekker Communication, PR, Social Media: Sjoerd Terborg Interactive Installation "Steve", Social Research: Joep Slagter Design Assistance, Realisation: Lion Zeegers Design Assistance: Marta Ardigo Actor: Steye van Dam Client: MU Eindhoven Completion Date: 2014 Photography: Corneel de Wilde

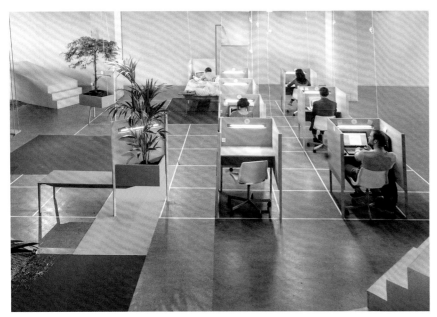

"走出办公室"象征着更好的合作和自由，符合 21 世纪的办公哲学。它提供了各种各样的办公场景，例如：有可爱白兔的花园、柔软舒适的"办公床"（还可以在等邮件的时候打个盹）、咖啡馆、客厅、图书馆。员工可以任意选择办公场所，设计营造出一种流动型办公空间。

在设计中，KNOL 工作室响应全新的办公潮流，在 300 平方米的开放空间内打造了一系列灵活而高效的办公环境，让员工们可以自由选择舒适的办公场景。"走出办公室"项目目前有五个特色办公区：大床、客厅、花园、图书馆和咖啡馆。设计师安娜称："我们不想设计一个完全没有办公氛围的环境（比如家居环境），也不希望办公环境里充斥着'工作'二字。我们想要一个流动的办公空间。"

习惯了格子间的员工反而对这种环境十分满意。"我们认为流动型办公是一件非常私人的事情。大多数人都喜欢在工作中享受自由。此外，他们很容易就能满足于灵活的环境和富有创造性的舒适设计。"设计师耶普解释道。设计师席琳·德瓦尔·马勒费吉特称："'走出办公室'是一个艺术项目，绝不是全面的科学实验。我们将人们带进了故事里。通过研究社交和建筑设计对办公效率和满意度的影响，我们对其有了更好的理解。"

设计师：KNOL 工作室（朱利安·科米林克、席琳·德瓦尔·马勒费吉特） 合作设计：克里斯蒂安·巴克尔 社会调查：安娜·戴克 传播、公共、社交媒体：舒尔德·特博格 交互装置"史蒂夫"、社会调查：耶普·斯拉格特尔 设计助理、实施：莱恩·齐格尔斯 设计助理：马尔塔·阿迪戈 演员：斯太尔·范戴蒙 委托方：埃因霍温 MU 公司 竣工时间：2014 年 摄影：科尼尔·德维尔德

Mobility

People Can Work Anywhere at Any Time

People can party late and sleep in, as long as they catch up on work later in the evening. People are their own boss. However, reports on the drawbacks of this freedom and flexibility in work are also proliferating. People are practically always at work. A lack of imposed discipline leaves people completely dependent on their self-discipline. Studio KNOL wants to investigate if people have gotten carried away with the trend of being extremely flexible. Would people enjoy having more set and limited working hours? Aren't people more productive when people add some externalised discipline to work environment? People considered it worthwhile to contribute to this debate by executing this experiment. The space underwent a physical and social transformation during which studio KNOL carefully monitored responses from Out of Office members.

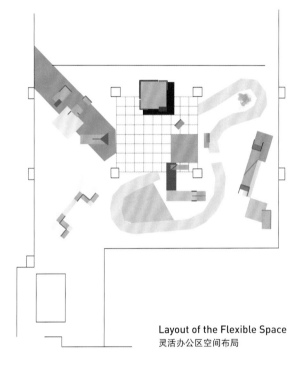

Layout of the Flexible Space
灵活办公区空间布局

人们可随时随地进行办公

人们可以派对狂欢到深夜，然后睡个懒觉，只要能在夜间完成工作就行。人们是自己的老板。然而，这种自由的缺点也很明显，它让人们几乎一直在工作中。缺乏强制性规定，只能完全依靠人们的自觉性。KNOL 工作室希望调查出人们是否会在极度灵活的状态下得意忘形。人们是否会更喜欢固定、有限的工作时间呢？办公环境的外界强制规定是否能增加人们的办公效率？这一切都值得进行项目的实验。经历了物理性和社会性改造之后，KNOL 工作室开始监控"走出办公室"项目的参与人员在工作中的行为。

Mobility

Cubicle Working Space

The five different atmospheres that were so characteristic for "Out of Office" gradually merged into a compact grid with 14 clear and ordered workstations. Interior architect, Christiaan Bakker, who joined the team, developed a smart system that made it possible to use the exact same building blocks for both set-ups. Workers can choose to do their own work in the cube and they don't have their own working space and everything is flowing. With these modular building blocks – three panels that together form one cubicle – the various work areas could easily be transformed into the office setting. The panels integrated lighting and electricity. All the grey-coloured building blocks for the cubicles form a clean grey space.

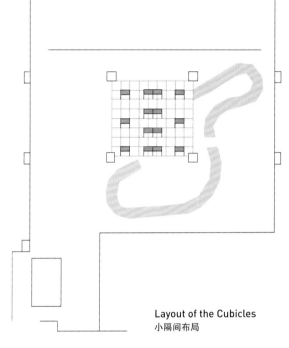

Layout of the Cubicles
小隔间布局

格子间办公环境

五种不同的氛围是"走出办公室"项目的特色，然而，办公空间里还有14个干净整洁的格子间。室内设计师克里斯蒂安·巴克尔开发了一个智能系统，让建筑模块得以实现了两种布局。员工可以选择在格子间里工作，没有独立的办公空间，一切都在流动中。格子间的办公模块由三块板材构成，实现了各种办公区域的快速转换。板材上嵌有照明和电力设施，形成了一个干净整洁的办公环境。

Mobility

FRIENDS OF EARTH
WORKSPACES DEFINED WITH SPECIFIC MISSIONS

London, UK

由具体工作选择不同的办公场所
地球之友 / 英国，伦敦

Recently the clients would like to adapt and transform their workplace into one that makes far better use of the workspace and functionality. That's why flow workspace becomes more and more popular. Flow workspace is a space to work chosen depending on the task in hand, encouraging more targeted and focused working as well as promoting movement and mobility throughout the workday.

To relocate 170 staff from their freehold Underwood Street office to the Printworks in Clapham, the office will provide a workspace for the whole organisation of 170 staff. However, this will only consist of 78 traditional workstations. The additional 92 seats will be made up of the quiet rooms, informal meeting spaces, acoustic meeting pods, touchdown benches and breakout spaces strategically placed throughout the office. It was an opportunity to visually link the organisation with their worthy causes for the planet and its resources and to constantly remind their visitors and staff of their aspirations for the organisation.

What if the plant is flat?
Travelling from the Northern to Southern Hemisphere, navigating through polar, middle and equatorial latitudes, is one aspect of this dramatic office experience. The other is recognising the potential of a flexible environment driven by contemporary ways of working rather than conventional methods.

Designer: peldonrose   Area: 930sqm   Completion Date: 2014

近来，越来越多的公司希望调整改造他们的办公空间，使其具有更好的功能性。因此，流动型办公空间越来越受欢迎。在流动型办公空间中，员工可根据手头工作选择办公场所，鼓励更多的定向集中办公，能够提升工作期间的移动性。

地球之友组织让170名员工从安德伍德街的办公室解放出来，来到克拉珀姆的新办公室。新办公室将为整个组织的170名员工提供办公空间，但是仅提供78个传统办公桌，额外的92个坐席将通过静室、非正式会议空间、隔音会议间、落地长凳、休息空间等呈现出来，遍布整个办公室的各个角落。这种设计在视觉上将地球之友组织与他们为保护地球及其资源所做的工作连接起来，反复向他们的访客和员工强调着该组织的雄心壮志。

如果地球是平的
办公室的设计带领人们从北半球到南半球，穿过南北极、中纬度和赤道。另一个设计特色是设计师以一种现代的办公方式打造了灵活的办公环境，而不是沿用传统的设计方法。

设计师：peldonrose 设计公司　面积：930平方米　竣工时间：2014年

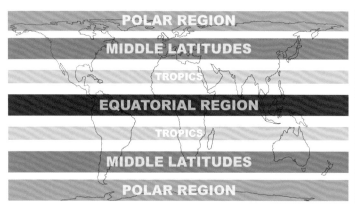

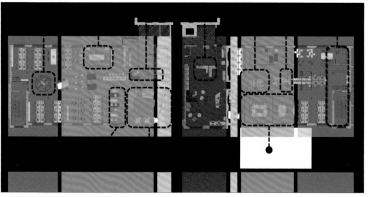

Dissecting the Earth in order to divide and zone the Workplace
To fully understand this project we must first understand how the spherical nature of the Earth can be represented in the seemingly flat plain of an office floor. As well as cracking this puzzle of planning, consideration should also be paid to the way the Earth is geographically divided and how latitude and longitude can be link with Friends of the Earth's brand identity to form a cohesive story.

解剖地球，划分办公区
为了全面了解这个项目，我们首先必须了解：地球的球形本质是如何被呈现在平坦的办公室楼面上的。在解决这个谜题之后，我们还要考虑地球的地理划分方式以及经纬度是如何与地球之友的品牌形象联系起来，构成完整的故事的。

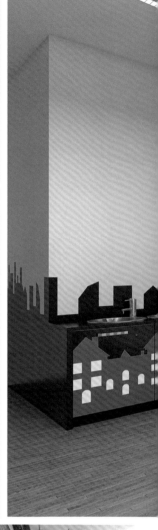

Equatorial Region Reception, Client Waiting and Tea Point
Red raises a room's energy level, stirs up excitement, draws people together and stimulates conversation. In an entryway it creates a strong first impression. Orange represents energy and a very social nature. It increases mental activity, appetite and energy. Hotter hues are the colours of welcome and warmth.
• Making sure all elements of geography are covered within the project, from the physical to the human, simple block shapes within the reception Tea Point are treated with graphic silhouette forms of cityscape scenes.
• Turn up the desert heat in Reception with bold red feature walls and warm sandy timbers.
• The recycled cardboard Reception Desk and cardboard lava stalactites created the brand DNA.

赤道区前台、客户等候区和茶水角
红色提升了空间的能量级,让人兴奋,把人们聚集起来,促进了对话。在入口处,红色打造了强烈的第一印象。橙色代表着能量和社交属性。它能增加智力活动,提升胃口和活力。暖色调给人以友好温馨的感觉。
• 保证所有地理元素都平铺在项目内,从物理特性到人文尺度。前台茶水角的简单积木造型就来自于城市风景的剪影
• 用鲜红的背景墙和温暖的木色来提升沙漠的热度
• 用回收纸板制成的前台和熔岩钟乳石体现了地球之友组织的根本

Open Plan Middle Latitudes South
Concept: Natural, Organic and Earthy

Exploring the construction of an office forest by wrapping existing cardboard columns in bespoke digital bark wall covering. A forest of bare trees marks the transition from rain forest biomes to the more desolate tundra of southern middle latitudes

采用开放布局的南半球中纬度地区
设计理念：自然，有机，朴素

把原有的纸板立柱用定制的数字树皮墙纸包裹起来，打造一片森林办公空间。一片光秃秃的树林标志着从热带雨林到更荒凉的南半球中纬度冻原地区的过渡。

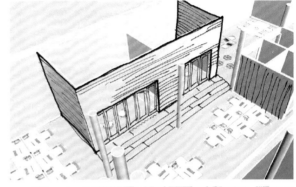

Mobility

Polar Region South
Concept: Cool, Professional and Crisp

Sitting in the shadow of the iceberg this open space team environment is a simplistic stripped-back space to work in. The Library flows around and through the work zone with a stand-up bench adjacent. Vinyl wall graphics are designed to cover the extreme back elevation of the office space with cool graphic images of towering ice formation in frozen seas. Ice white cubes of library shelving forms a calm corner in the open plan for quiet working and contemplative reference.

南极地区
设计理念：凉爽，专业，清新

以冰山为背景，这个开放的团队办公空间到处都显示出极简主义。图书室沿着工作区展开，并且通过竖起来的长凳穿过了工作区。乙烯基墙面图形的设计将办公空间的后墙面覆盖起来，用凉爽的海上冰山图案装饰着办公空间。书架上洁白的格子在开放式布局中围出了一个安静的办公角落，工人集中精神工作和沉思。

Polar Region North
White expresses purity, excellence, cleanness, clarity, simplicity and optimism. An increased sense of sophistication with a very clean feel.

What better place to cool down and clear your head than within an arctic iceberg? Two chat spaces are proposed within the crystalline dome of an iceberg-like structure that will be positioned in the centre of the North Polar Region of the office.

北极地区
白色象征着纯粹、优秀、干净、透明、简单和乐观,营造出非常简洁而精致的感觉。

如果要冷静下来、理清头绪,还有比北极圈的冰山更好的场所吗?两个冰山结构的圆顶小屋中分别是不同的聊天空间,它们就坐落在北极办公区的中央。

Staff Tea Point
Tropics
Protected from the open plan by an acoustic cave, the Staff Tea Point is an integrated extension of the workplace.

Grass Lands and Wet Lands
It's the perfect place to graze, to hide in the trees or perhaps drift in a boat.

员工茶水角
热带
隔音洞穴将员工茶水角与开放式办公空间隔开,这里是一个延展的办公空间。

草地与湿地
这里是放牧、在树林中藏身或划船漂流的好去处。

Open Plan Middle Latitudes North
Concept: Lush, Natural and Calm
Rolling grass hills, treehouses and boats come centre stage in this Northern biome. Depicting scenes of lush vegetation and plentiful water supply, it makes for the perfect setting for ideas to germinate and grow. Raised seating areas carved into hillside terraces form a perfect place to gather as a team, while acoustic seating pods offer cosy corners for smaller groups to huddle.

采用开放布局的北半球中纬度地区
设计概念：繁茂，自然，平静
起伏的草地山丘、树屋和小船构成了这个北半球生态系统的中央舞台。空间模仿了茂密的植被和充足的水源的场景，为灵感的滋生和成长提供了完美的背景。架高的座位区嵌入了山坡，形成了团队聚会的完美场所。隔音座椅则为小型团体提供了舒服的角落。

SKYPE'S NORTH AMERICAN HEADQUARTERS

DIFFERENT WORKSPACES FACILITATE DIFFERENT THINKING PROCESSES

Palo Alto, USA

通过不同的办公环境辅助不同的思维过程
Skype 公司北美总部 / 美国,帕洛阿尔托

A significant portion of Skype's culture is built around Scrum development (iterative idea generation) and a philosophy called "Agile Thinking" (the affect of environment on thought process). To support Scrum, Blitz designed a system of mobile white boards called Skype-its that are distributed throughout the project. They can be easily moved and stored depending on a development team's process and requirements. In this way, people can move to anyplace they want and increase the mobility of this working environment.

The building itself provided the greatest source of design inspiration. It was a dark and dingy space with years of tired tenant improvement projects layered on top of one another. It stood vacant for quite some time while neighbouring buildings were being leased within weeks. While touring the site, Blitz popped out a few ceiling tiles and the opportunity to strip away the layers, revealing the existing structure became immediately apparent. The architects made a decision to rip out all the existing ceilings and furring around the steel and never looked back. The resulting space is raw, industrial, and suggestive of a warehouse, which stands in perfect contrast to the highly refined meeting room pods that Blitz inserted into the open space.

With a tight budget, Design Blitz had to be creative and use simple, ordinary materials in new and creative ways. In the chill-out area they clad one wall with astroturf, and another with industrial felt. These coverings function as sound attenuation and finish materials, but also afford a little whimsy. At the Skype-it niches the architects used common 3/4" birch plywood and sized the niches to fit standard material modules. The plywood provides a high STC rating and is also the finish material. The Board Room is clad in wine flavour sticks, found at a salvage yard. These provide a unique finish to this particular pod and subtly celebrate California – an important programme element for Skype.

Designer: Blitz Area: 5,017sqm Completion Date: 2012 Photography: Hoffman Chrisman and Matthew Millman

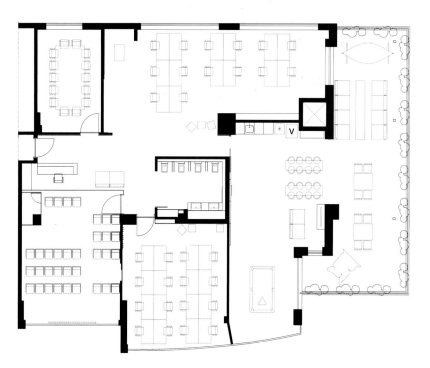

Mobility

Skype公司文化中极为重要的一部分围绕着迭代开发（迭代思想产生）和"敏捷思想"（环境对思维过程的影响）展开。为了辅助迭代开发，Blitz设计公司设计了一个移动白板系统并将其命名为Skype-it。这个系统遍布项目的各个角落，它们能根据开发团队的流程和需求方便地移动和储存。人们可以移动到任何位置，增强了办公环境的流动性。

建筑本身为设计师提供了极大的设计灵感。这是一个昏暗陈旧的空间，承租人的整修结果层层叠加。周围的建筑都能很快租出去，而这里却空闲了很久。在考察场地时，Blitz取出了几块吊顶板，剥离了层层装修，呈现出原始的建筑结构。设计师决定扯掉所有的吊顶和钢铁镶边。最终形成的空间充满了原始感和工业感，像一座货仓，与开放式空间里精致的会议室形成了完美的对比。

由于预算紧张，Blitz设计公司必须充分运用创新思维，创造性地运用简单、普通的材料。在休闲区，他们在一面墙上覆盖了人工草皮，另一面墙则覆盖了工业毛毡。这些覆盖物兼具隔音性和装饰性，同时又有些稀奇古怪。在Skype-it壁龛的设计中，设计师使用了普通的3/4英寸桦木胶合板，并且将壁龛设计成标准模块尺寸。胶合板和饰面材料具有极高的隔音等级。董事会议室包裹在酒红色的木板条里，就像救援修理厂。这种设计既形成了独特的装饰，又巧妙地赞美了加利福尼亚州，因为加州是Skype极为重要的程序元素之一。

设计师：Blitz设计公司　面积：5,017平方米　竣工时间：2012年　摄影：霍夫曼·克里斯曼、马修·米尔曼

With such large floor plates it was important for users to be able to orient themselves. The project is organised along a central highway with various nodal meeting spaces to convey a sense of location, while also encouraging workers to literally meet-in-the-middle.

由于楼面宽阔,因此用户的方向定位十分重要。项目沿着中央通道展开,两侧是各种节点会议空间,传递出一种方位感,同时也鼓励员工在办公室中间自由地会面交谈。

Mobility

Blitz also created a multitude of different environments to support different thought processes. All of the casual meeting areas are unique and mobile and there are three distinct phone booth types: light and bright for active thought; medium coloured for meditative thought; and dark cave-like rooms for introspective thought.

Skype required three distinct types of spaces: Mobility, Contemplation and Interaction spaces. Since Design Blitz believes that people, not conference rooms, deserve natural light and mobility, the architects took advantage of the exterior windows and located all workstations at the building perimeter (Interaction). They then worked their way back to the middle by degree of noise and distraction, with the noisiest functions being at the middle of the space (Mobility). Meeting rooms and phone booths were prioritized as no member of staff had a private office (an open floor plan with a benching workstation system was implemented). Contemplation spaces were interspersed in the form of overlapping casual lounges.

Blitz 还打造了各种不同的环境来辅助不同的思维过程。所有的非正式过程都保持了独特性和流动性，共有三种类型的单间设计：浅色而明亮的活跃思维空间、中性色彩的沉思思维空间、深色的洞穴式反省思维空间。

Skype 要求这三种空间具有不同的特征：流动性、沉思性、互动性。由于 Blitz 设计公司坚信人比会议室更需要自然采光和流动性，建筑师充分利用了外层窗户，把所有工作台都设在建筑的外围（互动性）。噪声最大、干扰最大的功能区被设在了空间中央（流动性）。会议室和单间进行了优先设计，因为员工们都没有私人办公室（项目选择了开放式办公布局，引入了长凳办公台系统）。沉思空间以叠加的休闲室的形式点缀在各处。

Mobility

The pods house the meeting and collaboration function and require a very high level of acoustic attenuation (far exceeding typical TI construction). To achieve the aesthetic of a floating pod (the pods are self-supporting) and achieve Skype's acoustical requirements, Blitz had to develop a unique building typology. The architects utilised a structural roof deck, commonly used in large span construction, which provided them with the structural diaphragm over the pods, provided a high level of acoustic attenuation, and also functioned as the finished ceiling. This multi-function quality was a key focus throughout the project – one item fulfilling multiple functions.

承担着会议和协作的单间要求装饰材料有极高的隔音等级。为了实现美观性和隔音性的双重要求，Blitz必须开发出一套独特的建筑设计。设计师利用大跨度工程中常用的结构屋顶板为单间的天花板提供了结构隔板，实现了高等级的隔音效果，同时也兼具装饰吊顶的功能。这种多功能属性是项目设计的关键——一个物品实现多重功能。

PART 5

SMALL AND SMART – HOW TO MAKE FULL USE OF WORKSPACE

第 5 章 集约型办公空间——
如何实现"空间尽其用"

SMALL OFFICE: WHEN EVERY INCH COUNTS

With progressing nations and cluttered lives of people, today, the office spaces are outgrowing their conventional definitions. The increasing health consciousness and environment awareness have paved way for incorporation of leisure zones, breakout areas, fitness rooms, informal lounges in client requirements. With smaller offices getting into the scenario, it is the moral responsibility of designers to come up with more responsive, functional and suitable design ideas.

Today, where modern man spends majority time at his work place, our offices need to be communicative, technologically advanced and green conscious – in all "smart". Space should be efficiently zoned out into independent activities to facilitate proper functioning in a small office. Small does not imply that the spaces need to get cluttered. It is essential to give each activity its due area and not integrate functions into one. A simplified zoning must be highlighted to make the space more responsive and make the most out of the given space.

Walls can be done away with. Partitions can be limited to a lower height and incorporated as a design element. Glass walls, available in innumerable variants can be used in an artistic manner and made to blend with the overall space. We need to get rid of the traditional matchbox concept of workstations and adopt more open layouts where there is no barrier between spaces. As a result, our space is more efficient for the user, facilitates visual connectivity, promotes transparency between colleagues and provides better interaction and communication.

Technology needs to be weaved appropriately in the design of the office to provide a more comfortable environment for the user. Use of technology such as Wi-Fi, video conferencing, plasma screens and biometric systems is making offices more efficient. Using automation can be a small step towards the smart etiquette. Automation can reduce the manual work load on a small space, making services easier and even add on to the green points. The modern man is blessed to have achieved so many technological advancements, and we as designers should exploit them to the most by amalgamating them in our design to make the office a more appropriate and customised place.

小型办公空间：寸土寸金

随着国家的进步和人类生活的多样性，今天，办公空间已经超越了它们的传统定义。人们与日俱增的健康意识和环境意识让办公设计增加了休闲区、休息区、健身房、娱乐室等多种要求。小型办公室日渐兴起，而设计师们有义务提供更负责、更实用、更适合的设计方案。

现代人的大多数时间都消耗在工作场所，我们的办公室必须便于交流，拥有先进的技术，有绿色环保意识，总而言之，要"智能"。小型办公空间的分区必须高效合理，以辅助各项独立活动的有效进行。每项活动都要有自己的空间，不能简单地合并。必须突出简化的分区设计，让空间更具适应性，充分利用已有的空间。

墙壁可以拆除。隔断可被限制在较低的高度，与设计元素相结合。各种各样的玻璃墙可经过艺术处理与整体空间融合起来。我们必须抛弃传统的火柴盒式工作台设计，采用更开放的布局，使空间之间没有阻碍。最终，我们的空间会让用户感到更高效，形成更好的视觉连通性，建立员工之间的透明关系，提供更好的互动和交流。

办公空间的设计必须恰当地考虑技术需求，为用户提供更舒适的环境。无线网络、电视会议、等离子显示器、生物识别等技术的应用让办公空间变得更高效。自动化技术是迈向智能时代的一小步。自动化技术能削减手工作业的负担，让服务变得更便捷，设置还可以提升环保指数。现代人已经获得了如此多的技术进步，作为设计师，我们应当在设计中充分利用它们，把办公室变成更合适的定制空间。

办公室还应当是一个更健康的办公空间。空间的自然采光和阳光直射显得至关重要。此外，自然通风也是一个重要的方面。各个空间都必须有充足的采光来保证正常运行。合适的空气循环对小型办公空间来说是最基本的元素。办公区内必须点缀绿色空间，它们可以融入设计，兼作能促进用户交流

The office should also be conceptualised as a healthier workplace. Emphasis should be on infusing natural light and direct sun rays within the space. Natural ventilation is an essential aspect. Care must be taken to cater adequate light for proper functioning of each space. Proper air circulation is an essential for a small office space. Effort must be made to permeate green spaces within the office area. Such spaces blend in the design and double up as an interactive space or breakout space where users can exchange ideas. A lot of innovative materials that are eco-friendly can be experimented with in a small space to do justice to the smart aspect.

A great office doesn't need to be large; in fact, there are plenty of benefits to being scrappy that large companies try to imitate. And while it's important to be smart about reducing clutter and finding storage alternatives for a small space, designing a truly great small office requires more insight and planning than a few Murphy tables and IKEA storage bins.

Here are a few tips on how to design a small office that will help facilitate communication, bring in business, and keep the employees happy.

Designing a Small Office: Use a Layout that Reflects Your Organisational Structure

How are decisions made in your business? All companies, and especially small firms, must be clear about more than just their budget before they can start designing their offices. The clients need to consider the flow of internal direction when designing the office. It's not only just about their business model, but how it's how they plan on running their business. How do they plan on communicating from within? Is it top down? Does the president make decisions and people follow? Or is it a collaborative environment where people work together? How does the space incorporate everyone so that the employees feel like they're getting vested in the company?

Even if you're dealing with a space 500 square feet or smaller, the positioning of each employees matters. Is the CEO situated in the corner with a big office, or stationed on a desk with the interns? The placement of employees in a confined area reflects the flow of direction of the organisation. For example, a manager who manages six employees, sits right in the middle of his office. It gives him a

的互动空间或休息空间。小空间的设计可以尝试许多创新环保材料，以实现更智能的设计。

一个优秀的办公室并不需要多大；事实上，大公司也试图模仿小型办公空间的生机勃勃。由于小空间的设计必须消除杂乱感并寻找替代的存储空间，一个优秀的小型办公空间需要更好的眼光和规划，不仅仅是几张办公桌、几个收纳箱那么简单。

以下是一些小型办公空间的设计技巧，它们能帮助公司营造更好的交流氛围，带来商业机会，让员工更快乐。

小型办公空间设计：用布局设计来反映你的组织结构
你的公司如何做出决定？所有公司，特别是小公司，必须对自己办公室设计的预算、布局等做出详细规划。在办公室的设计中，客户需要考虑到内部的流线。这不仅关乎他们的商业模式，还关乎他们计划如何运营公司。如何实现内部交流？由上而下？是董事长做决定，下面的人执行？还是在协作环境中让大家一起工作？如何利用空间让员工们融入公司，使他们获得归属感？

即使你的空间只有不足50平方米甚至更小，每个员工的位置也都很重要。首席执行官是坐在大办公室的角落，还是坐在实习生的旁边？员工的位置安排反映了整个组织的流向。例如，如果管理者需要管理6名员工，他的座位设在办公室的中央。这样一来，他就能看清每个人都在干什么，而他的员工也能很方便地接近他。

小型办公空间设计：反映公司的态度
如果你需要会见客户，你的办公室就必须精准反映你的商业目标。在公司空间的设计中，设计师应当设法了解客户对公司的看法、公司的品牌策略以及公众对公司的看法。他应当提问："公司想呈现出什么样的形象？"

即使你的公司不需要会见客户，你也必须考虑到当前以及未

chance to see what everyone is working on, and it gives his employees ample opportunity to approach him.

Designing a Small Office: Reflect Your Company's Attitude
If you see clients, your office should be an accurate reflection of your business objectives. When designing spaces for companies, the designer tries to understand how customers view the firm. He learns about their branding and he tries and sees how their business is reflected to the public. He asks: "What is the image this company wants to represent?"

Even if your company doesn't see clients, you must ask yourself how your office space represents itself to current and future employees. Because the architecture reflects an attitude: how you choose to practice, how you choose to communicate, and depending on what type of firm it is, how you place your values.

Designing a Small Office: Use Frugality to Your Benefit
If all the furniture the company owns was used, mostly bought on Craigslist or left over from a previous tenant, what was important was to have a space that was quirky, fun, and colourful. When people sit down in a more relaxed atmosphere it helps them be more thoughtful. They could go out and buy 10 Herman Miller chairs and a bunch of nice tables, but that could be in any office in the world.

Why buy from an office catalog? With a small office, you have more latitude to be frugal by buying second-hand.

"What I love is the fact that you move from your desk to an area that feels different, and an area that lends itself to creative thinking," the clients say. "We're not talking beanbag chairs. We just found the right balance."

Designing a Small Office: Create Small Perks for a Big Impact
There's an interesting theme there, which is you want people to feel like they're living a bit of the "good life" to be able to have espresso and beer in frosted mugs. Those little things go a long way.

Certainly, an important issue to address is whether or not you should spend money on perks like these, or if you should you just pay people more. Having a

来员工对办公空间的印象。建筑反映态度：你选择如何经营、如何沟通以及公司的核心价值是什么。

小型办公空间设计：节约的好处
如果公司的办公家具全都是二手的，大多数来自于二手网站或是前一家承租人留下的，那么设计的重点就在于如何使空间看起来奇妙有趣又多姿多彩。当人坐在一个放松的环境中，他更能深刻思考问题。公司可以买10把办公座椅和一堆新的办公桌，但是与世界上任何一家公司并不会有什么不同。

为什么要买二手办公用品？对小公司来说，二手用品能节约不少开支。

"我喜欢离开办公桌到另一个区域，在一个完全不同的地方进行创造性思考"，客户说，"我们并不需要变形椅，而是想找到正确的平衡点。"

设计小型办公空间：小福利，大惊喜
这个主题十分有趣。浓咖啡、冰啤酒等小福利能让人感到自己生活在幸福之中，这些小细节有着深远的作用。

当然，问题的重点在于你是否应当花钱来提供这些小福利，还是直接给员工更多的工资。在小办公室里，咖啡、啤酒等福利更加简单，无需花费过多的钱财和精力。但是设计师会指出，这些小福利在行为经济学上有着重大的意义。如果老板说："我每天多给你两美元去买咖啡"，其效果肯定不如在办公室里放一台咖啡机。你希望人们热爱办公室。如果晚上7点能让我喝一杯冰啤酒，我将获得极大的满足。

"小而智能"并不仅是一句空话，而是需要时间来实现。是时候超越常规体制，聚焦于崭新的设计理念了。随着电子设备的智能化，我们也该为小型办公室注入智能设计，使其能够面对与大型空间相同的挑战。我们需要设计出能留住人的办公空间，因为他们没有更好的工作场所了。人们前往一个优秀的办公空间的原因是他们想去，而不是必须去。

small office gives you an opportunity to offer things like coffee and beer because it's not that expensive to replenish and maintain. But the designer may point out interesting facet of behavioural economics in this situation that make the perks seem worth it. If the boss says, "I'll give you two bucks a day so you can go get coffee," that's not as valuable as having an espresso machine in the office. You want people to love this place. At 7 p.m. when I can have a frosted beer – it's a huge draw.

"Small and Smart" is not just a slogan, but the need of the hour. It is the time to look beyond the normal regime and focus on more novel design philosophies. As the gadgets get smart, it's time for us to do the same with smart offices that are small yet capable of meeting the challenges of a large space with same might. We need to go ahead and create spaces where people would want to come every day because there is no better space to do their work. People come to a great workplace not because they have to but because they want to.

Further ahead is an assortment of such impeccable small office spaces which will reflect the innovations and responses by creative people around the world. We hope this will inspire you to create "Small and Smart" office spaces.

Kapil Aggarwal
President of Spaces Architects@ka

接下来的案例汇集了各种出色的小型办公空间，展示了来自全球各地的创意设计，希望这对想要设计"小而智能"办公空间的人士有所帮助。

卡皮尔·阿加沃尔
Spaces Architects@ka 公司总裁

ARCHITECTURE STUDIO AND COWORKING SPACE OFFICE

THE CONTRAST BETWEEN DARK GREY AND WHITE ENLARGES THE SPACE VISUALLY

Ferrol, Spain

深灰与纯白的对比产生了放大空间的效果
建筑工作室与协作办公空间 / 西班牙、费罗尔

Recently the size of offices is not so much important. Designers would like to build their "cabin" as a shelter capable of making themselves feel comfortable while they develop their work as architects and also make the office enlarged.

With this project, the designers tried to renovate a typical ground floor space in a building inside Ferrol's 18th century's historical centre of a narrow and elongated plot (5×16m), with a few sunlight hours. In order to do enlarge this office, they created an "upside down boat keel"; a long and white structure that works as a space where they can take cover and where their work takes place.

Specification and Materials:
• Red pine framework, white colour painted, dimensions: 100×70mm
• Waterproof white laquered MDF forming the "cabin", dimensions of planks: 19×150mm
• Laminated floor FINSA, 90H SOBERANE ARTIC OAK
• OSB 16mm board panels coating walls of the rest area

Designer: As-Built Arquitectura, Interiorismo, Infografía   Client: Espacio As-Built   Area: 82sqm   Completion Date: 2014   Photography: ©Moncho Rey

Sections of the Space
空间剖面

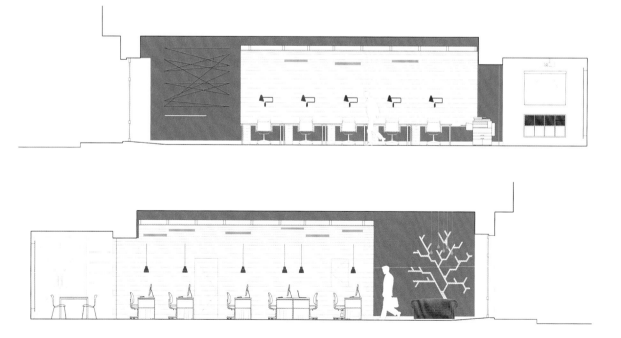

近年来，办公室的规模已经不再那么重要。设计师喜欢建造自己的"小屋"作为"庇护所"，那里能让他们感到轻松舒适。他们为自己建造办公室，同时还能进行扩建。

项目对西班牙费罗尔18世纪的老城中心内的一个典型的底层楼面空间进行了翻修改造。整个空间十分狭长（5米X16米），每天的日照时间有几个小时。为了扩大办公室，设计师打造了一个"翻转的船龙骨"：白色的狭长结构构成了一个空间，为他们提供了庇护所，也是他们的办公场所。

材料规格：
- 红松框架，白漆，尺寸：100×70mm
- 防水白漆中密度纤维板建成"小屋"，木板尺寸：19×150mm
- 叠层地板FINSA, 90H SOBERANE ARTIC OAK型
- 16mm定向刨花板，覆盖休息区的墙面

设计师：As-Built建筑事务所、Interiorismo设计公司、Infografía设计公司　委托方：As-Built空间设计公司　面积：82平方米　竣工时间：2014年　摄影：蒙乔·雷伊

Small and Smart

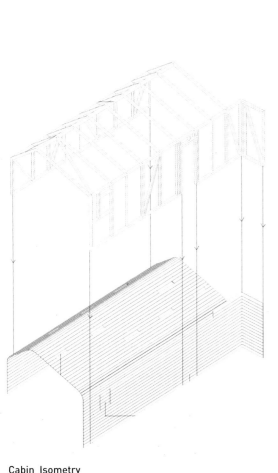

Cabin Isometry
"小屋"等距图

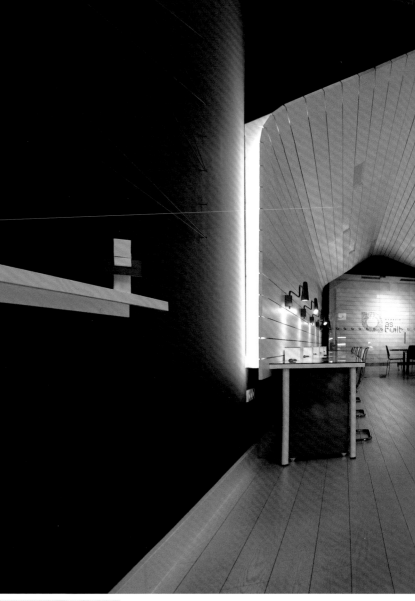

The designers designed this cabin so that their "keel" doesn't reach all the way to the façade, creating a lobby that does the job of receiving their clients as well as a place to gaze and admire the cabin. It is here where a tree-shape sculpture created with white plastic cups welcomes them. With this Tree, they want to make reference to man's primary shelter against Nature, which as time passes will turn into the cabin that they are trying to represent. It also makes the workers feel like working at the grand and bright space.

小屋的设计让设计师的"船龙骨"不会延展到立面上，形成了一个用于接待客户、欣赏办公室的空间。由白色塑料杯组成的树形雕塑迎接着来来往往的人们。这棵树象征着人类最初的庇护所，随着时间的流逝，它逐渐演变成了我们的小屋。整体设计让员工们感觉自己置身于一个宏大而明亮的空间。

Details of the Wall
墙壁细部图

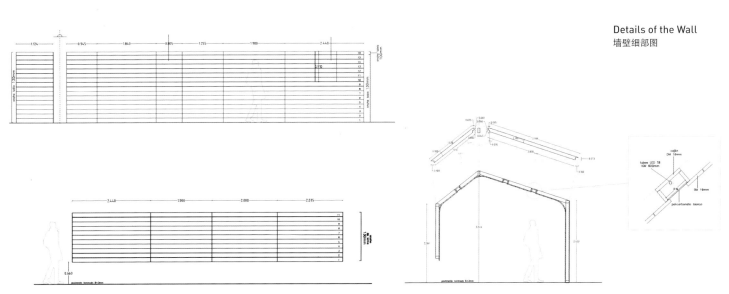

Small and Smart

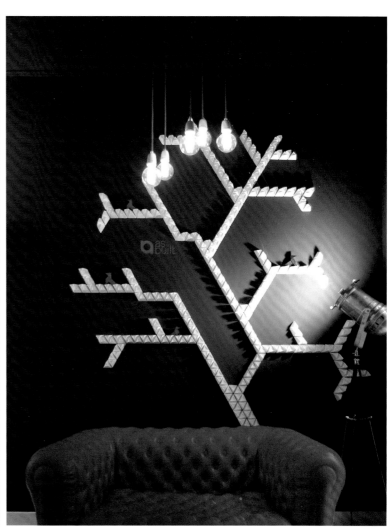

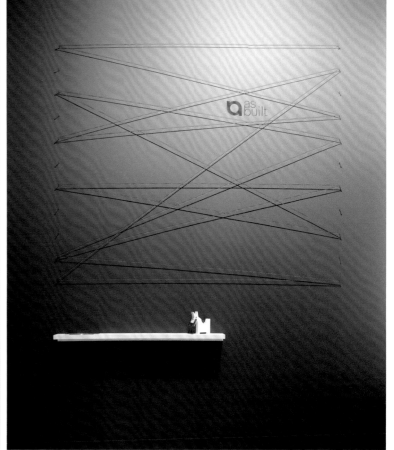

All the construction was made by dry joint, with a system similar to the American's "balloon frame", due to a framework of red pine wood (70×100mm.) covered (and braced) with white lacquered MDF strips (19×150mm.).

Once you are outside the "shell" of the cabin, the rest-area provides a small kitchen and storage selves. Here the walls were coated with a warmer material, OSB panels (16mm), in contrast with the dark grey walls and ceiling.

所有结构都采用干缝连接,采用了与美国"轻型木构架"相似的系统:红松木框架(70×100mm)上覆盖着漆成白色的中密度纤维板条(19×150mm)。

走出小屋的"外壳",休息区里是一个小厨房和储藏架。这里的墙壁上覆盖着质感更温暖的材料——定向刨花板(16mm),与深灰色的墙壁和天花板形成了对比。

The whole office (walls and ceiling) was painted dark grey in order to contrast with the stark whiteness of the floor and the lacquered MDF surfaces of the cabin. The contrast helps enlarge the whole working space.

Thanks to indirect LED lighting, the initially dark space was transformed into a bright warm ambient that, consequently, cuts down on energetic use.

整个办公室(包括墙壁和天花板)都被漆成了深灰色,与纯白的地板和小屋的白色中密度纤维板形成了对比。这种对比有助于放大整个办公空间的视觉效果。

间接LED照明让昏暗的空间变得明亮而温馨,同时也大幅缩减了能源消耗。

FLUID FORMS – CUBIC OFFICE

FLUID FORMS AMPLIFY THE SPACE

New Delhi, India

流线的运用产生了扩大空间的作用
流动造型——立方办公室／印度，新德里

An interesting ceiling design can liven up the environment and create a certain feel in the office. Not necessarily always first on the list of priorities when it comes to interior fit-outs, ceilings can expand the scope of interior and enlarge the space.

The office for real estate consultant has been conceptualised as a modern white office with fluid forms. The client requirement was to have two MD cabins with Conference for eight seaters with a reception and waiting. The design concept was intended to try and experiment with fluid forms. Also it was a challenge to create a small but complete and white interior space. The design and concept was very rare for the designers' style of working, and the execution involved lot of site visits which involved redesigning and refining.

Designer: Spaces Architects@ka   Client: Cubix Homes
Area: 111sqm   Completion Date: 2014   Photographer: Bharat Aggarwal

有趣的天花板设计能活跃环境，在办公室营造出某种特定的氛围。在室内装修中，天花板也许并不被列在第一位，但是良好的天花板设计却能拓展室内视野，扩大整个空间。

这个为置业顾问公司所设计的办公室以现代的白色空间和流畅的造型为特色。委托方要求设计两间独立的小屋作为会议室和经理办公室，并配备8个座椅和接待等候区。设计概念的目标是利用流动造型进行试验，打造一个小而完整的白色室内空间。项目的设计和概念在办公空间中都十分独特，其实施需要经历多次现场考察，且涉及了多次修改和完善。

设计师：Spaces建筑事务所   委托方：Cubix置业公司   面积：111平方米   竣工时间：2014年   摄影：巴拉特·阿加瓦尔

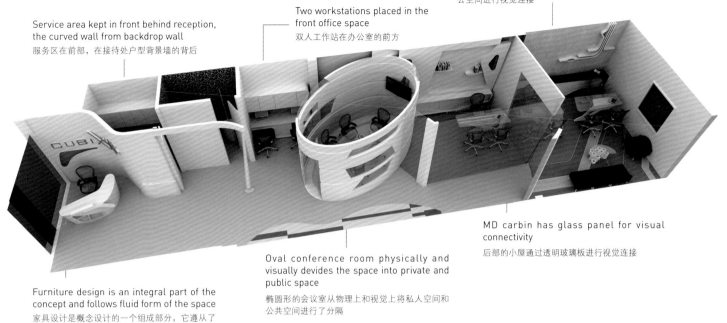

Service area kept in front behind reception, the curved wall from backdrop wall
服务区在前部,在接待处户型背景墙的背后

Two workstations placed in the front office space
双人工作站在办公室的前方

Rear MD cabin has dark flooring reflecting ceiling profile. The design intension is to visually connect the office space through glass panels
后部的小屋用黑色地板反衬天花板的轮廓。设计意在通过玻璃板将办公空间进行视觉连接

Furniture design is an integral part of the concept and follows fluid form of the space
家具设计是概念设计的一个组成部分,它遵从了空间的流体形式

Oval conference room physically and visually devides the space into private and public space
椭圆形的会议室从物理上和视觉上将私人空间和公共空间进行了分隔

MD carbin has glass panel for visual connectivity
后部的小屋通过透明玻璃板进行视觉连接

Small and Smart

The service areas are kept in front behind the reception. The front of the office being 11'(3.4m) with kitchen wall in fluid profile forms a backdrop for the reception. A building model designed has been placed vertically on the wall in front of reception adding character to the space. The reception table designed following the fluid concept of the space with abstract backlit panels is in harmony with curved back wall which takes a peel form, the top curve panel extending towards workstation and other supported by angled column. The ceiling in the front area has abstract forms with the one above the reception having multiple battens with semi-elliptical backlit panel. All the fluid and curve lines help enlarge the office space.

服务区被设在前台背后。办公室的前端宽 3.4 米，流线造型的厨房墙壁形成了前台的背景墙。前台正前方墙壁上的建筑模型为空间带来了特色。前台接待桌的设计沿用了空间的流动概念，配有抽象的背光照明板，与弧形背景墙和谐相融。墙壁上采用"剥落"设计，顶部的弧形板向工作台延伸，而其他的板材则由角柱支撑。办公室前端的天花板采用抽象造型，前台头顶的天花板由多个板条构成，嵌有一个椭圆形的背光板。各种流畅的曲线都有助于扩大办公空间。

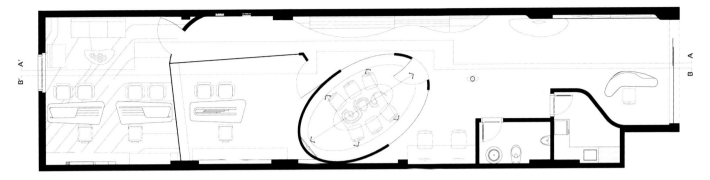

Floor Plan
楼层平面图

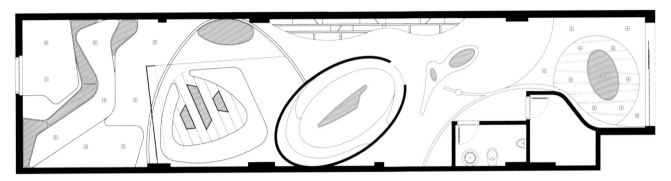

Ceiling Plan
顶棚平面图

The conference room has multiple layered panels with glass slit. The form at the centre acts as a transition dividing the office in public and semi-private spaces. The fluid extends to the ceiling and to the hanging light also. The conference table has been designed by combining multiple curved panels fixed together with a glass top in elliptical shape. The floor has backlit glass floor creating an ambience in the space. The corridor leading to the rear room has been designed with multiple project images display in black which with fluid form ceiling in abstract shape backlit panel reflecting the pattern on the floor also. The cabin behind the conference has angled glass panels on either side to visually connect with the outer space. The glass panels are intersected by curved partition which extends from the rear room to the corridor extending to the conference room partition supporting the glass door below. The ceiling of the cabin designed in curved profile with multiple grooves has an abstract-shape ceiling hanged below. The fluid and abstract form extends to the furniture designed in the space. The rear MD room has two MD tables with sofa, the flooring has black tile in contrast to the grey used outside to create transition, and the ceiling has been designed in fluid form with abstract-shaped black painted panels backlit. The MD table has multiple glass panels fixed together with abstract-shape panels.

会议室有多层挡板，上面嵌有玻璃开口。它坐落在空间正中，起到了公共空间与半私人空间的隔断和过渡作用。流动感一直向上延伸到天花板和照明设施上。会议桌的设计融合了多块弧形板，上面是一个椭圆形的玻璃桌面。背光式照明的玻璃地面营造出一种太空的氛围。通往后部房间的走廊两侧展示着各种各样的黑白建筑图片，头顶的流线形抽象背光板与地面遥相呼应。会议室后方的小屋两侧都是玻璃板，与外界形成了视觉联系。玻璃板被弧形隔断隔开，后者一直从房间延续到走廊，拓展到会议室的隔断处。小屋的天花板采用弧线设计，由多个凹槽构成了抽象造型。后方的经理办公室配有两张办公桌和沙发，地面采用黑色地砖，与外面的灰色地砖形成对比和色彩的过渡。造型流畅的天花板上配有黑漆背光板进行装饰。办公桌由多块玻璃板拼接起来，造型抽象。

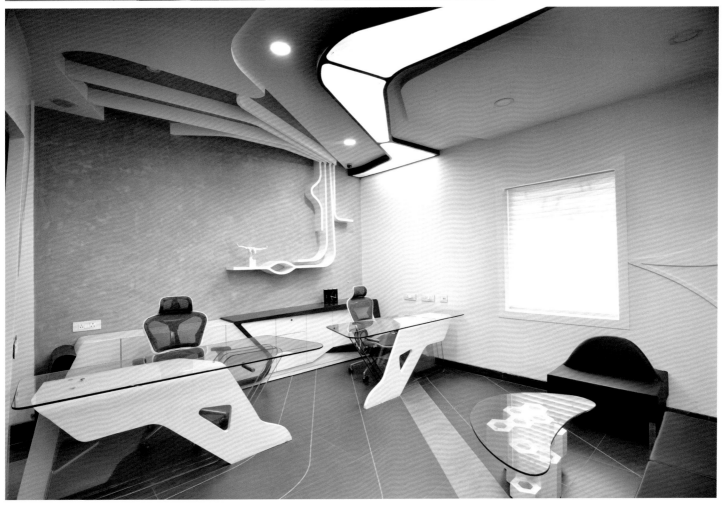

SISII SHOWROOM AND OFFICE IN KOBE

THE SMART USE OF PLANTS PROMOTES THE OFFICE'S QUALITY

Kobe, Japan

巧妙利用植栽提升办公空间品质

Sisii 公司神户展示厅兼办公室 / 日本，神户

Today's technology has enabled us to work anytime and anywhere. Performance is no longer determined by time spent on the office but by results. The trend is that less individual cubicles exist and more often we see companies encouraging open spaces that allow greater interaction and communication.

In this case, the designers designed a living office bringing spaces back to life, creating places that brought people together again. The goal: to have more efficient workers; more productive hours spent at work, and to enhance a better quality of life. Satisfied with the results of the previous project Yuko Nagayama & Associates did, the client chose to take this same experience and apply it to their new offices: a firm ready to take its workers to the next level: Interaction and Communication.

Designer: Yuko Nagayama, Yuko Nagayama & Associates
Area: 144sqm Completion Date: 2012 Materials: Phosphoric acid zinc plate

Small and Smart

当今的技术让我们能随时随地进行办公。办公效率不再由办公室的工作时间决定,而是取决于结果。在这种趋势下,小隔间越来越少,公司更加倡导开放空间的设计,以实现更好的互动和交流。

在本案中,设计师所设计的生活办公空间让空间回归生活,把人们重新聚集起来。设计的目标是让员工的办公效率更高,使他们在办公室里的工作时间更加高效,并且提升他们的生活品质。委托方对永山裕子建筑设计事务所早先的项目十分满意,因此决定在新办公室中委托他们进行设计。公司已经做好了准备,决定将员工带到下一个更高的层次:互动和交流。

设计师:永山裕子、永山裕子建筑设计事务所　面积:144平方米
竣工时间:2012年　材料:磷酸锌板

Floor Plan
楼层平面图

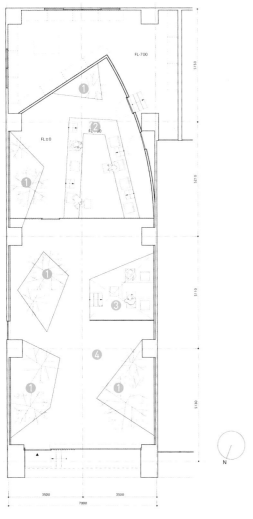

1. Planting space　1. 绿色植物
2. Desk space　　 2. 带桌办公区
3. Meeting space　3. 会议空间
4. Showroom　　 4. 展示空间

The mission was to provide employees with an alternate space where work and relaxing could exist. Informal working areas, a cafeteria to interact, open library where they could work with each other or with customers. To work while doing something nice like having a cup of coffee, a work-lunch while listening to music near the green plants.

设计的目标是为员工提供一个兼具工作和放松氛围的交替空间。非正式办公区、互动餐厅、开放图书室都是员工办公或接待顾客的理想场所。他们可以边工作边享用咖啡和工作午餐,或是在绿色植物旁听音乐。

Section 空间剖面

Small and Smart

The designers consider living vegetation an indispensable element to create an interactive and relaxing environment. A suspended iron sheet with wrought holes emulates the connection to relaxing area and office. Beneath these iron sheets plants are grown to represent the outer Rokko Mountains habitat and reveal nature through these openings. The iron sheets are partially peeled and hoisted, and turned into a meeting space, and in the office into a large desk.

On the walls between the pillars and beams, mirrors are placed in places in the existing building, so that inner space expands in various directions. In order to create a space agile, the designers created gently curved walls inside the house. With those tactically set mirrors and curved walls, we do not see the dead end in the office space.

Mr. Toshiya Ogino, garden designer and the collaborator, actually living in this area of Rokko, created indoor gardens with a concept to emulate an original relaxing environment. The planting job was proceeded while confirming the whole picture including the perspective reflected in mirrors. The trees and the artificial hills made of lava were placed carefully confirming the balance of reflection image.

As a solution to indoor planting problems, the designers adopted the method combining the use of inorganic lightweight soil and appropriate drainage and ventilation system. In addition, indoor grow light is turned on during the night-time to give the plants enough lighting for photosynthesis.

设计师认为绿植是营造交互式放松环境所不可或缺的元素。一块带有开口的悬挂式铁板起到了休闲区和办公区之间的连接作用。铁板下方的植物代表着六甲山栖息地，展示了自然的风光。铁板的一部分被剥开升起，变成了会议空间和巨大的办公桌。

在梁柱之间的墙壁上安装着镜子，从各个方向实现了室内空间的视觉延伸。为了营造灵活的空间，设计师在室内打造了具有柔和曲线的墙壁。精心搭配的镜子和弧形墙壁将办公空间的尽头巧妙地隐藏起来。

园艺设计师荻野俊哉就住在六甲山地区，他承担了室内花园的全部设计，为办公空间打造了别出心裁的轻松环境。植栽的设计工作考虑了全局效果，包括镜子中映出的景象。树木和熔岩制成的假山的设计与镜中形象形成了和谐统一的效果。

为了解决室内种植问题，设计师选择了无机轻质土，并配置了合适的排水和通风系统。此外，办公室的室内生长灯在夜间也会打开，为植物提供充足的照明，以实现光合作用。

The hanger racks in the showroom are placed around the plants so that they have a look of tree branches. This fashion brand is well-known for leather products and most of products are in earth tone colours, so hanged garments look like leaves and pupas.

Transforming roles accordingly to the spots, the iron sheet extends, as neutral platform coexisting with the vital energy of the vegetations will enable an active site for informational exchange.

Seen from the frontal road, customers seeing the products in the showroom, people having a meeting, and staff working in the shop – they all look as actors performing one scene on stage. Across the steel panel, a piece of nature of Rokko appears under the view of the scene. It is believed that this scene itself expresses the mind and concept of the brand.

展示厅的衣架环绕着植物摆放,就像树枝的一部分一样。Sisii 公司的品牌以皮革制品为主,大多数产品都呈大地色系,因此,悬挂的服装看起来就像树叶和蚕蛹。

铁板在不同的位置扮演了不同的角色,它与生机勃勃的植物共存,为人们提供了一个相互交流的平台。

从建筑前方的道路上,顾客能看到产品陈列以及公司人员开会和办公的景象——他们就像舞台上的演员,上演着一幕幕的戏剧。透过铁板,六甲山的一角显露出来。设计师坚信这些场景能够体现品牌的精神和理念。

从多种意义上来讲,集装箱是一种理想的建筑材料。它们坚韧、耐用、成本价格低、环保,因此深受欢迎。由集装箱建成的办公室又小又智能,就像一个完整的圆圈,十分完美。

黄金比例建筑办公室是对小型智能空间的一次尝试。宽阔的木板墙面形成了一扇迎接人们进入公司内部空间的大门。代表着公司成员的三顶安全帽与金属螺丝构成的公司LOGO共同标志出办公空间的入口。"黄金分割"带领着人们进入黄金比例的世界。

设计师:Xrysitomi Collective建筑事务所　竣工时间:2013年　摄影:德米·卡拉萨菲里

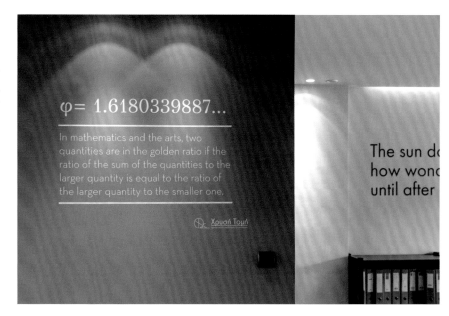

GOLDEN RATIO HEADQUARTERS

THE SLOPING ROOF AND OPENINGS OVERCOME THE LIMITS OF SMALL SPACE

Athens, Greece

斜屋顶与开窗克服小空间的局限性
黄金比例总部 / 希腊、雅典

Containers are in many ways an ideal building material. They are welcomed become of their strength, durability, low cost and eco-friendliness. Such offices made of containers are small and smart just like a complete circle.

The Golden Ratio architectural office is an alternative approach to the concept of small and smart. A broad wooden surface forms a welcoming gate to the inner space of the company. Three construction helmets, as many as the company members, along with the imposing logo made of metal screws, mark the entrance to the working space. "Two quantities are in the golden ratio if ..." and journey into the world of Golden Ratio begins.

Designer: Xrysitomi Collective Architects　Completion Date: 2013　Photography: Demi Karatzaferi

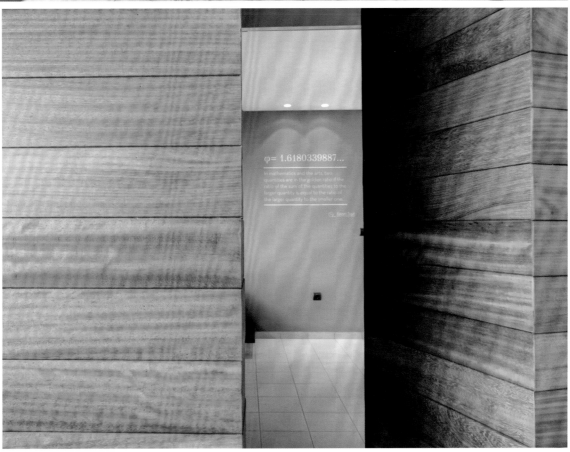

Small and Smart

1st Floor Plan
2层平面图

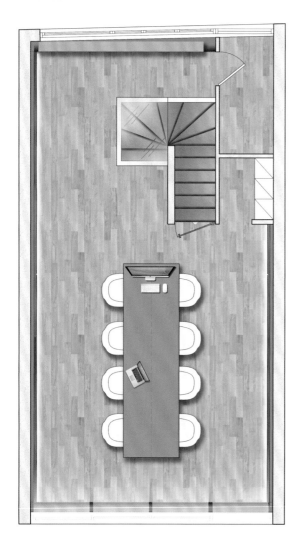

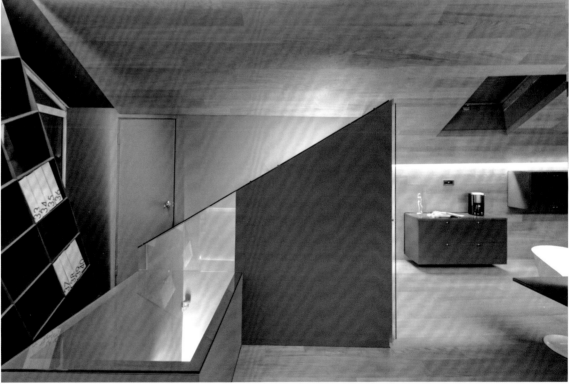

Following an upward course, the visitor reaches the specially designed shell, open to all kinds of activities and to collaboration between different disciplines. A single wooden surface covers the walls and the sloping roof forming a closed core, which shields the customer relationship with the company by creating a climate of confidence and intimacy and also overcome the limitation of small size. The openings on both sides, as well as those on the roof, allow natural light to invade space, making the office enlarged, expressing the evolution and inspiration, characteristics of the Golden Ratio.

一路向上,访客将到达精心设计的集装箱外壳,开放式空间鼓励各种活动和跨领域的合作。单块木平面覆盖着墙壁,斜屋顶形成了封闭的内核,在消费者和公司之间营造出一种相互信任的亲切氛围,同时又克服了小空间的局限性。两侧墙壁和屋顶的开窗引入了自然光,在视觉上放大了空间,表现了黄金比例的特点——进化和启发。

Scetion
空间剖面

On one side of the room light is filtered by the metal construction of the library, which through the change of grid results in subversive design. The elongated anthracite metal workspace hosts the meetings, while at the same time emphasises the geometry of space. Lacquer cabinets of similar colour discreetly complement the composition. The result is characterised by simple and dynamic lines, expressing a design stable and sincere, like the relationship of the company with customers.

房间一侧的光线透过图书室的金属结构渗透进来，而格栅的变化则形成了颠覆性的设计。细长的炭黑色金属办公空间是会议室，其设计突出了空间的几何造型。相近色彩的橱柜形成了谨慎的空间组合。简洁而动感的线条体现了设计的稳定性和真诚感，就像公司与消费者的关系一样。

Small and Smart

AFTERWORD
后记

Colourful surfaces, various materials, warm textures, natural and artificial light. It is not a playground nor a fancy café to hang around; it is the contemporary perception of what a working space looks like.

Motivating, interesting and mind-blowing, the new working spaces are here to stay. In an attempt to inspire employees to work productively and feel better at the same time, companies are putting all their effort to create working spaces that any design freak would be jealous of. Trying to make people feel comfortable, stay longer and be more relaxed, creating a cosy and home-like atmosphere is often the goal of contemporary office-design process.

Varying from pop-art versions of graphic design to elegant interpretations of space, creative office design can be as diverse as it gets. Boring rows of identical desks and piles of paperwork belong to the past; now it is the time to boost the opportunities of office layout design. Thematic corners, cosy spots and recreation spaces are the trend, organising and categorising space.

"The whole idea here is that by having an open floor plan where people work close to each other, it facilitates people sharing and communicating about what they're doing, which enables better collaboration, which we think is key to building the best services for our community"; these words belong to no other, than Mark Zuckerberg, explaining the culture of Google office design. Wide office floor with an open plan, without heavy walls dividing the teams is a design solution selected by many companies to facilitate effective communication at work.

"Form follows function" as it is, but why not, function follows form as well? Form can acquire a dominant role in design, even if it is about the working environment. Sensual curves and dynamic lines, exploding volumes and continuous surfaces, form can determine everything.

In the past, the desired effect of an office space was to keep you concentrated by constantly reminding that you are at work. Nowadays, companies seem to put into implementation the idea that "this is not a working space". The key is to feel so comfortable, forget that you are at work, so that you can work better.

Contemporary office design uses a variety of means to be effective. Space layout, forms, materials, textures, colours are equally important to produce an interesting

色彩斑斓的墙面和地面，各种各样的材料，温暖的质感，舒适的自然光和人造光。这不是游乐场或时尚的咖啡厅，这是人们对现代办公空间的预期。

富有启发性、有趣、让人兴奋，这些新办公空间深受欢迎。为了激励员工努力工作、提升他们的舒适度，企业正倾尽全力打造连设计狂人也嫉妒的办公空间。为了让人们感到舒适、待得更久、更放松，大多数现代办公设计都以家一般的舒适氛围为目标。

无论是流行艺术风格的平面设计，还是优雅的空间诠释，创意办公空间的风格多种多样。一排排无聊的同质化办公桌和成堆的文书属于过去，现在的办公布局有更多的选择。主题角、休闲区、娱乐空间越来越流行，它们将空间分类组织起来。

马克·扎克伯格在解释谷歌的办公文化时说道："这里的整体概念是打造一个让人们可以亲密工作的开放空间，它有助于人们的分享和交流，能促进更好的合作，而合作则是我们为社区提供更好服务的关键。"许多公司都选择宽敞的办公楼层和开放的布局来促进工作中的有效交流。

"形式追随功能"，确实如此，但是为什么不让功能也追随形式呢？即使是在办公环境里，形式在设计中所扮演的角色也不容小觑。感性的弧线和活泼的线条，爆炸的空间和连续的表面，形式能够决定一切。

过去，办公空间的理想效果是让人专注集中，不断提醒你在工作。现在，企业似乎更倾向于传递"这不是一个工作空间"的概念。重点是舒适，忘了你在工作，才能更好地工作。

现代办公设计利用各种方式提高效率。空间布局、形式、材料、材质、色彩在打造趣味办公环境的过程中同样重要。挑出条条框框来思考是一大技巧。无论是室内花园、秋千办公椅，还是床式办公桌，结果都是一样的：放松和创新。

另一个重要的因素是识别度。公司的梦想、目标、哲学和原则在办公设计的体现上已经变得前所未有的重要。办公空间

working environment. Thinking outside the box is definitely the clue. Whether it is an indoor garden, a swing-office chair or a bed-like desk, the result is the same: relax and be creative.

Another important factor is identity. Company's dream, goals, philosophy and principles reflect on office design now, more than ever. Working space is a part of the company branding; it is an opportunity to show who you are and what you do. It is in fact what makes Skype or LinkedIn or Google be themselves. Have a strong identity, but not be afraid to be different; that is the point.

In a constantly changing world, office creative design follows the pace for sure. In this book, some of the most inspiring working space examples from all around the world are being presented. From Tokyo to Athens, from industrial simplicity to rustic atmosphere, from minimalistic white spaces to graffiti-covered conference rooms, they are all here. Sit tight and enjoy!

Margarita Skiada
Fellow Architect of Golden Ratio Collective Architecture

体现了一部分的公司品牌形象，是展示公司的一大机会，正如Skype、领英、谷歌一样。具有强烈的企业形象，不惧不同，才是重点。

世界不断变化，创意办公设计必须随之而变。本书汇集了来自全球各地的极富启发性的办公空间设计案例。从东京到雅典，从简约的工业风到淳朴的乡村风，从极简白色空间到涂满涂鸦的会议室，应有尽有。好好享受阅读的乐趣吧！

玛格丽塔·齐亚达
Golden Ratio Collective Architecture 建筑事务所资深建筑师

INDEX
索引

1305 STUDIO
www.d1305.com
Room 210,
No. 383, Changhua Road,
Shanghai, China
Tel: +86 21 62995615
Fax: +86 21 62998175

As-Built Arquitectura Interiorismo Infografía
www.as-built.es
Moncho Rey Lage
arquitecto colegiado COAG nº 3.440
Tel: +34 633 719 783
Pablo Ríos Sánchez
arquitecto de interiores
Tel: +34 620 811 330

AP+I Design
www.apidesign.com
AP+I Design Inc.
117 Easy St.
Mountain View, CA 94043

Apostrophy's The Synthesis Server Co., Ltd.
www.apostrophys.com
290/214 Ladprao 84 (Sungkomsongkraotai 1) Wangtonglang
Bangkok 10310 Thailand
Tel: +02 193 9144
Fax: +02 193 9143

Bean Buro Architects
www.beanburo.com
Unit 2803, 28/F, Sunlight Tower, 248 Queen's Road East,
Wan Chai, Hong Kong
Tel: +852 2690 9550

BroekBakema
www.broekbakema.nl
Van Nellefabriek
Schiehal G
Van Nelleweg 1
3044 BC Rotterdam
The Netherlands
Tel: +010 413 47 80
Fax: +010 413 64 54

Camenzind Evolution
www.camenzindevolution.com
Samariterstrasse 5
8032 Zurich
Switzerland
Tel: +41 44 253 95 00

Design Blitz
www.designblitzsf.com
435 Janckson ST, San Francisco, CA 94111
Tel: +1 415 525 9179

Designliga
http://en.designliga.com
A. Döhring, A. Stanojčić GbR
Hans-Preißinger-Str. 8, Halle A
D-81379 Munich
Germany
Tel: +49 0 89 624 219 40
Fax: +49 0 89 550 697 97

gfra architecture
gfra.gr
Komninon 37, 11473 Athens, Greece
Tel: +30 210 6427972
Fax: +30 210 6427203

Golden Ratio Collective Architects
www.xrysitomi.gr
4 Aiantos str., Vrilisia, 15235
Tel: +30 210 6131174
Fax: +30210 6132196

i29 interior architects
www.i29.nl
Industrieweg 29
1115 AD Duivendrecht
The Netherlands
Tel: +31 20 695 61 20

IwamotoScott Architecture
www.iwamotoscott.com
729 Tennessee Street
San Francisco, CA 94107
Tel: +415 643 7773 Main
      +415 864 2868 Mobile

Kapil Aggarwal
www.spacesarchitects-ka.com
A-21/A, Basement, South Extension-II

New Delhi - 110049
Tel: +011 26268108/09

Klein Dytham architecture
www.klein-dytham.com
1-15-7 Hiro
Shibuya
Tokyo 150-0012
Tel: +81 3 5795 2277

Liong Lie Architects
www.lionglie.com
Sint-Jobsweg 30
3024 EJ Rotterdam
The Netherlands
Tel: +31 0 10 478 2064

NARUSE INOKUMA Architects
www.narukuma.com
Tel/Fax: +03 6915 1288

PARAT
www.buero-parat.de
Juliusstraße 12
22769 Hamburg
Tel/Fex: +49 40 413040 43

Peldonrose
www.peldonrose.com
Sterling House, 42 Worple Road, London, SW19 4EQ
Tel: +020 8971 7777

Setter Architects
22 He-amal St,
Ramat Gan 52572
Israel
Tel: +972 3 613 7771
Fax: +972 3 613 7774

Spaces Architects@ka
www.spacesarchitects-ka.com
A-21/A, Basement, South Extension-II
New Delhi - 110049
Tel: +011 26268108/09

Studio Niels™
www.studioniels.nl
Stokstraat 59

6211 GC Maastricht
Tel: +31 43 326 1001

studio KNOL
www.studioknol.com
Berberisstraat 16, 1032 EL
Amsterdam, The Netherlands
Tel: +31 6 34148155

Studio Yaron Tal
www.yarontal.com
Hasharon st. 12 Tel Aviv 66185
P.O.Box 1057 Tel Aviv 61010 Israel
Tel: +03 6881113
Fax: +03 6872887

teamLab Architects
http://architects.team-lab.com

Threefold Architects
www.threefoldarchitects.com
Threefold Architects
Great Western Studios,
Studio 203,
65 Alfred Road ,
London,
W2 5EU
Tel: +020 8969 2323
Fax: +020 7504 8704

Virserius Studio
http://virseriusstudio.com
150 W 28th Street, Suite 1704
New York, NY 10001
Tel: +1 212 337 2195

Yuko Nagayama & Associates
www.yukonagayama.co.jp
Yuko Nagayama & Associates
1-39-1-4F, Momoi, Suginamiku,
Tokyo, Japan 167-0034
Tel: +81 3 6913 7097
Fax: +81 3 6913 7098

图书在版编目（CIP）数据

办公空间创意设计 /（英）德里斯科尔编；

常文心译 . -- 沈阳 : 辽宁科学技术出版社 , 2016.6

ISBN 978-7-5381-8078-7

Ⅰ . ①办… Ⅱ . ①德… ②常… Ⅲ . ①办公室 - 室内装饰设计 Ⅳ . ① TU243

中国版本图书馆 CIP 数据核字 (2016) 第 030725 号

出版发行： 辽宁科学技术出版社
（地址：沈阳市和平区十一纬路 25 号 邮编：110003）
印 刷 者： 上海利丰雅高印刷（东莞）有限公司
经 销 者： 各地新华书店
幅面尺寸： 215mm×285mm
印　　张： 17
插　　页： 4
字　　数： 50 千字
出版时间： 2016 年 6 月第 1 版
印刷时间： 2016 年 6 月第 1 次印刷
责任编辑： 杜丙旭
封面设计： 周　洁
版式设计： 周　洁
责任校对： 周　文

书　　号： ISBN 978-7-5381-8078-7
定　　价： 298.00 元

联系电话：024-23284360
邮购热线：024-23284502
http://www.lnkj.com.cn